Scottish
Wild
Plants

THEIR HISTORY

ECOLOGY

AND

CONSERVATION

Scottish Wild Plants

THEIR HISTORY
ECOLOGY
AND
CONSERVATION

Philip Lusby and Jenny Wright

with photography by Sidney J Clarke

Edited by

Norma M Gregory

ROYAL
BOTANIC
GARDEN
EDINBURGH

EDINBURGH: THE STATIONERY OFFICE LIMITED

© Crown Copyright 1996
First Published 1996

ISBN 0 11 495802 5

Text and Photographs Copyright Royal Botanic Garden Edinburgh

Frontispiece: *Saxifraga oppositifolia*

Phil Lusby is responsible for the Scottish Rare Plant Project, which is based at the Royal Botanic Garden Edinburgh.

Jenny Wright is currently in her final year of a postgraduate study investigating the conservation ecology of whorled solomon's-seal, *Polygonatum verticillatum*.

Sidney Clarke is Principal Photographer at the Royal Botanic Garden Edinburgh.

Applications for reproduction should be made to
The Stationery Office Limited.

British Library Cataloguing in Publication Data.
A catalogue record for this book is available from the British Library

Contents

Acknowledgements

It is a pleasure to acknowledge the guidance and encouragement of David Long and Henry Noltie who read and commented on the manuscript. We are also very grateful to Peter Pitkin for his helpful comments on the introductory text.

For information and advice on individual species we thank the following: Dr Adrian Dyer, Roy Harris, Dr Peter Hollingsworth, Dr A.C. Jermy, Lyndsey Kinnes, Dr Stuart Lindsay, Dr Barry Meatyard, R. Desmond Meikle, Dr Gordon Miller, C. W. Murray, Dr Adrian Newton, Prof. Jim Parks, Dr Peter Pitkin, Gordon Rothero, Dr Mark Watson, Dr David Welch and Dr Chris Wilcock.

For help with localities and assistance with photography we thank: Donald Bruce, John Burlison, Dr Neil Cowie, Dr Jim Dickson, David Ellis, Dr Chris Ferreira, Clare Geddes, Bill Helyar, Mr & Mrs La Croix, Dr Stuart Lindsay, Ron McBeath, Iain MacDonald, Murdo MacDonald, Neil McIntyre, Margaret Maan, David Mardon, John Mitchell, C. W. Murray, Alex Scott, Dr Chris Sydes, Barry Unwin, Robert Unwin, Peter Wortham and Chris Wright.

We are also extremely grateful to Mary Bates for her illustrations and to all the Library staff for their unfailing help.

During the course of writing this book, The Scottish Rare Plant Project has been largely funded by the Headley Trust, to whom we are sincerely grateful.

Foreword

The native flora of the British Isles has been the subject of a remarkable and most welcome upsurge of public interest in recent years. This interest has been fuelled by an increasing realization that the habitats of many of our native plants have already been lost or are under serious threat. Within Scotland, however, we still have an astonishingly varied range of plant communities, although there is little room for complacency. Some of them, the mires of the southern uplands for example, are extremely fragile; others, although more robust and extensive, are nevertheless dwindling under the many pressures on land use. A concern to preserve our remaining natural and semi-natural habitats, and to replace some of our lost native vegetation, has generated many local and national conservation schemes, and is the inspiration for the Scottish Rare Plant Project. This important initiative is based at the Royal Botanic Garden Edinburgh and since its foundation has received additional support from Scottish Natural Heritage, the Headley Trust, the Hamamelis Trust, the Imlay Foundation, the Robertson-Ness Trust, the Highland Society of London and Johnston Press, as well as the Trustees of the Royal Botanic Garden Edinburgh. The Garden thanks all these donors. *Scottish Wild Plants* has been produced by two members of the Project team to provide knowledge and understanding of the plants and the habitats they are working to conserve.

We are fortunate that in the year of publication a number of excellent conservation initiatives have received support from the Millennium Fund. Among these are measures to restore and extend the Caledonian pinewoods, the home of several of the plants described here. This ecosystem, a major vegetation type a 1000 years ago, has been largely

reduced to a few remnants in remote glens – a fitting
subject indeed for a Millennium restoration project.
Success in the conservation of our native flora, however,
whether in the great wood of Caledon or the arctic-alpine
refuges of the high tops, or the machair that fringes the
lonely shorelands of the far west, depends on a thorough
knowledge of all its components. Sadly we are far from this
ideal, but I hope that this book will make a significant
contribution to the wider understanding of what is already
known and what still remains to be done. I hope that it will
also be seen as a celebration of the beauty and diversity of
the Scottish flora.

The authors have selected a range of species for inclusion in
the book: some are common but nonetheless worthy of
attention; others are rarities that, without informed
conservation measures, could well vanish from our land.
Some of the species selected are large – for example, the
long-lived Scots pine that has provided timber since man
first built houses and boats in Scotland – others are small
– such as the tiny *Primula scotica* whose flowers grace
northern cliffs for a few months each year. All, however, are
integral components of our natural heritage and that of our
children and grandchildren. If biodiversity, that much used
but rarely defined concept, means anything, we must
recognize and cherish all the plants and places described
here, and the many others that could not be included
because of the constraints of space. This concern for our
native flora should not be a matter merely for scientists, but
must be the concern of us all. Therefore, even though the
book has been prepared by professional botanists, great
care has been taken to extend its appeal to all who have an
interest in and a love of Scottish plants. I am sure that it is
equally worthy of a place on the bookshelves of land
managers, hill-walkers and all who love the countryside as
on the desks of plant scientists and conservationists.

The Nature of the Scottish Flora

HISTORICAL BACKGROUND

The distribution of plants is never static. All species respond to changes in the environment and our present flora results from millions of years of interaction between plants and the environment, but increasingly modified by human influences within the last 3000 years.

Whether any plants definitely survived the height of the last ice age some 15,000 years ago still remains questionable. Recent research on Scots pine (*Pinus sylvestris*) and other species suggest that a few plants may have been able to hang on in isolated pockets, but it is more likely that after the ice age virtually all the flora recolonized from outwith the glaciated area which extended to southern England.

The withdrawal of the main ice sheet left an ice-scoured landscape in Scotland similar to that in Greenland today. Water was locked up in ice, lowering the sea level, and frost and melt-water very much influenced the colonizing vegetation.

In slightly warmer conditions ice-free ground became colonized by arctic vegetation with abundant heather (*Calluna vulgaris*) and other related shrubs interspersed with large areas of grasses, sedges, and flowering herbs. The soils were young and had not been subjected to long periods of heavy rain that washes or leaches out minerals. Consequently, they were generally more basic than today and the fossil record indicates this in the abundance of mountain avens (*Dryas octopetala*). Other arctic plants more widespread in the Late-glacial but very rare today are Iceland purslane (*Koenigia islandica*) and Norwegian mugwort (*Artemisia norvegica*). Gradually, woody plants such as juniper (*Juniperus communis*) appeared and the first tree, birch (*Betula*), colonized. The climate improved erratically, and for a short period became colder again, marked by another increase in *Dryas*. After this set-back the climate began to warm progressively and this is considered to mark the beginning of the true Post-glacial period, about 10,000 years ago. Lakes formed in hollows and provided a habitat for aquatic plants whilst the open vegetation was colonized by a sparse forest of birch. Over low ground pine ousted the birch and grassy heaths. The Boreal period began with increasing warmth allowing birch to spread northwards and to higher altitudes. Hazel (*Corylus avellana*) became quite widespread and occasionally formed thickets under an open canopy of birch and pine. Pine continued to spread in the warmer and drier climate and reached its maximum cover in this period. At the same time elm and oak began to increase in southern Scotland.

Around 7,000 years ago the Atlantic period began, marked by a continuous rise in temperature and significant increase in rainfall. Ground water levels rose and minerals were leached from the soils making them more acid. Breakdown of dead plant remains was retarded under these conditions and accumulation of peat resulted in the formation of bogs. The wetter climate also created conditions favourable for the moisture-loving alder (*Alnus glutinosa*). This warm, wet period is known as the Post-glacial Climatic Optimum. Britain now became separated from the European continent. The circulation of the sea increased the humidity and oceanicity of the British climate especially in relation to that of continental Europe. Warmth-loving (thermophilous) species spread at the expense of the vegetation of earlier colder periods which became confined to cooler conditions at higher altitudes. Oak continued to spread from the south, became dominant in parts of the south and west Highlands and spread along the coasts of east and west Sutherland reaching the Inner Hebrides. Pine was better adapted to the open glacial sands of the eastern Highlands, although substantial tracts of pinewood developed in the west. Highland woods were generally dominated by pine, juniper and birch with alder in wetter places, but on base-rich soils over more calcareous rocks, mixed ash and elm woodland developed. The treeline reached a slightly higher altitude during the Atlantic period than today. Plants of open habitats other than those of bogs were largely confined to areas of thin soil where trees could not colonize, such as mountains (above the treeline), screes, coastal fringes and rocky outcrops.

A further climatic change some 5,000 years ago brought cooler, drier conditions and with this commenced the Sub-Boreal period. The surface of Scottish bogs dried sufficiently to allow pine to spread, except in the extreme north where birch, willow and juniper kept their ground. In north-east Sutherland and Caithness trees did not establish in any quantity and this area became dominated by blanket bog with patterned pool systems (the Flow Country). The cooler conditions lowered the altitudinal treeline slightly and allowed montane plants to expand their territory.

Wetter conditions marked the opening of the Sub-Atlantic period about 2,000 years ago and allowed the resumption of bog growth at the expense of forest cover which steadily declined. Birch generally appears to have been the dominant tree overall throughout the Post-glacial period but there were marked regional differences. During the Sub-Atlantic the effects of man first become really noticeable as forest clearance, cultivation of crops and grazing increased. Human influences will be discussed later but, leaving these aside, the natural composition of our native vegetation and flora has been shaped largely by changes in climate since the last ice-age.

GEOGRAPHICAL ELEMENTS

The present composition of the British flora largely reflects climatic conditions which have prevailed for the last 2,000 years. Because nearly all our plants have immigrated from the European continent, their distribution outside Britain provides an indication of the geographical affinities of our native flora.

The British plant geographer J. R. Matthews identified 15 groups of plants which reflected distributional patterns he termed geographical elements. As the British Isles are situated on the north-west margin of Europe and their north–south axis spans about 12° latitude with most of the uplands in Scotland, one might expect latitudinal differences in representation of the various geographical elements. Over half the British flora belongs to three of Matthews' elements: the general European Element (species with a wide European distribution but absent from Asia); the Eurasian Element (species widely distributed in Europe and also occurring in Asia); and the Northern Hemisphere species which occur in Europe, Asia and North America. The geographical elements particularly well represented in Scotland are the arctic-subarctic (species absent from central Europe) and the arctic-alpine (often circumpolar species with their main area of distribution in the Arctic but also occurring at higher altitudes in the mountains of Europe). An important but small group of species form the Alpine Element which are those from the central European mountains but absent from northern Europe and the Arctic. Few of our arctic-subarctic or arctic-alpine species are rare in Europe or Scandinavia, most having larger populations near the centre of their ranges. A notable exception is *Artemisia norvegica* which only occurs in central and southern Norway and the Ural mountains in Russia.

Except in a few localities in Wigtownshire and Kirkcudbrightshire, Scots lovage (Ligusticum scoticum) replaces the southern rock samphire (Crithmum maritimum) in maritime cliff communities around the Scottish coast.

In terms of species number it is well known that the British flora is not rich and this is generally explained by the obstacle to immigration formed by the North Sea when Britain separated from the continent in Atlantic times. However, because of Britain's geographical position, some unique assemblages of plants from different geographical elements occur. One of the most fascinating is the combination of the arctic-subarctic *Koenigia islandica* with the alpine cyphel (*Minuartia sedoides*) on the Trotternish hills on Skye and the alpine purple oxytropis (*Oxytropis halleri*) with the arctic-alpine alpine cinquefoil (*Potentilla crantzii*) on Ben Vrackie in Perthshire. On the east and west coasts of Scotland the arctic-subarctic

3

11 = Lammermuir Hills
12 = Pentland Hills
13 = Renfrew Hills
14 = Campsie Fells
15 = Glen Farg
16 = Ochil Hills
17 = Sidlaw Hills
18 = Ben Laoigh
19 = Breadalbane Hills
20 = Ben Lawers
21 = Rannoch Moor
22 = Ben Vrackie
23 = Glen Clova
24 = Caenlochan Glen
25 = Glen Doll
26 = Ardnamurchan Peninsula
27 = Loch Shiel
28 = Creag an Dail Bheag
29 = Cairngorms
30 = Glen Feshie
31 = Loch Insh
32 = Rothiemurchus
33 = Abernethy
34 = Loch Kishorn
35 = Banff
36 = Buckie
37 = Nairn
38 = Ben Eighe
39 = Shieldaig
40 = Ben Dearg
41 = Inverlael Forest
42 = Freevater Forest
43 = Seana Bhraigh
44 = Elphin
45 = Inchnadamph
46 = Flow Country
47 = Durness
48 = Cape Wrath
49 = Islay
50 = Colonsay
51 = Ardmeanach
52 = Mull
53 = Sound of Mull
54 = Coll
55 = Barra
56 = Benbecula
57 = North Uist
58 = Skye
59 = Trotternish Hills
60 = Hoy
61 = Old Man of Hoy
62 = Unst
63 = Keen of Hamar

1 = Cairnsmore of Fleet
2 = Merrick
3 = Rhinns of Kells
4 = Silver Flowe
5 = Kirkconnel Flow
6 = Cheviot Hills
7 = Mull of Kintyre
8 = Whitlaw Mosses
9 = Carrick Hills
10 = Moorfoot Hills

SUF = Southern Upland Fault
HBF = Highland Boundary Fault
GGF = Great Glen Fault

Scots lovage (*Ligusticum scoticum*) meets rock samphire (*Crithmum maritimum*) which belongs to the Oceanic Southern Element consisting of species occurring mainly in southern and western Europe including the Mediterranean. A few plants such as pipewort (*Eriocaulon aquaticum*) are confined to a few localities on the western Scottish mainland and the Hebrides. These have their main headquarters in North America and are extremely restricted in Europe. How this peculiar distribution developed has puzzled plant geographers for many years. The species composing these unique combinations are at their geographical limits where, because of marginal environmental conditions, evolution is likely to be most active. This situation may manifest itself in different ways. For example the alpine catchfly (*Lychnis alpina*) is restricted to soils with a high metal content at its two British localities (Angus and Cumbria) whereas in Scandinavia it grows on a wider range of soil types. The snow gentian (*Gentiana nivalis*) is confined to montane calcareous grassland and ungrazed ledges in Scotland, but again in Scandanavia it is able to colonize lowland habitats. Thyme broomrape (*Orobanche alba*) is a very variable plant in southern Europe ranging from white to deep red, but the form that reaches Scotland is particularly uniform in colour.

Among our few endemic species the beautiful Scottish primrose (*Primula scotica*) and Shetland mouse-ear (*Cerastium nigrescens*) are two of the better known examples. Other variants have been classified as varieties or subspecies and others may yet warrant taxonomic distinction after further research.

Besides individual species, some plant communities and vegetation types in Scotland are internationally important. The oceanic climate favours heather (*Calluna*) and the large tracts of heathland in the eastern Highlands managed for grouse shooting are exceptional. The high altitude prostrate form of this heathland, often with abundant lichens and bryophytes, is rare outside Scotland. Our damp oceanic climate compared with that of Europe is especially favourable for the development of blanket bog. This type of peatland forms on level or gently sloping ground, and the living bog surface is maintained by high rainfall rather than by groundwater. The Flow Country of Caithness and Sutherland is one of the finest examples of this habitat in the world.

An example of a uniquely Scottish vegetation type is the machair grassland which is particularly well developed in the Outer Hebrides. This is formed by the deposition of calcareous wind-blown shell sand driven on-shore by westerly gales. The turf is short and springy and is typically devoid of marram grass (*Ammophila arenaria*) which is more confined to sand dunes. Traditionally managed machair, under a regime of low intensity grazing with no use of chemical fertilisers, produces a veritable display of flowering herbs in summer.

REGIONAL FEATURES

The influence of climate on the distribution of plants and vegetation types can be clearly seen in the preceding remarks. A secondary factor is the nature of the rock and soil. Variation in these results in marked regional differences in plant life in otherwise comparable localities.

The climate of Scotland is dominated by the combined effects of the warm western Gulf Stream and its rain-bearing air flow. In comparison with the climate of continental Europe, that of Scotland is mild, wet and windy. The average temperature generally declines from south to north and west to east. On low ground the most favoured area is the southern Outer Hebrides which on average remains free from air frost from the end of March to the beginning of January. In contrast, the glens of the Eastern Highlands remain free of air frost from mid-June to mid-August only. Temperature decreases with altitude and on the hills of the central and eastern Highlands air frost may occur in any month. Rainfall increases from east to west (and also with altitude) and the area which receives the most rainfall in Scotland is a band some miles inland stretching from Ross-shire to Argyll. Areas with low rainfall are the east coast lowlands of the Moray Firth, East Lothian and Berwickshire. These are broad generalizations and local topography can modify regional climatic conditions. For example, in mountainous terrain the increased prevalence of mist and cloud reduces daily bright sunshine which, in turn, reduces average temperatures. Wind also modifies temperature and is strongest at the coast. The windiest places in Britain, the Outer Hebrides and Shetland, have more than thirty gale days per year.

The raised mires of south-west Scotland and north-west England support strong populations of Bog rosemary (Andromeda polifolia).

The rocks and soils of Scotland are predominantly acid. The acidity of soils is generally increased by the high rainfall which leaches out nutrients, and consequently much of the vegetation in Scotland is acidophilous. Areas of lime-rich or calcareous rocks form more fertile soils and support a greater variety of plants. These areas are botanically important. Drainage from calcareous rocks may sometimes enrich acidic soils, a process called flushing. Sometimes this occurs on such a small scale that the only indication is the presence of a few calcicolous (lime-loving) species within an otherwise acidic flora. Hard calcareous rocks such as gabbro may give rise to quite acidic vegetation because the rock is resistant to weathering and does not readily yield a supply of nutrients. Conversely some easily-weathered soft acidic rocks may form a more mineral-rich soil and allow more nutrient-demanding plants to grow.

Southern Scotland

This region extends from the English Border to the Southern Upland Fault which crosses Scotland from Dunbar in the east to Loch Ryan in the west. The predominant rocks are Ordovician and Silurian sediments which form the Southern

Uplands. Granitic intrusions in the south-west create a more rugged topography and include hills composed of metamorphosed sediment and granite such as the Merrick, Rhinns of Kells and Cairnsmore of Fleet. In the east the main area of high ground is the Lammermuir and Moorfoot hills with the Cheviots forming the border. Mires are a feature of this area. Examples of fine basin mires which are enclosed in glacial water-logged depressions over calcareous sediments occur near Selkirk (the Whitlaw Mosses). These depressions contain a range of base-rich swamp, mire and tall-herb communities with birch-willow carr. Cranberry (*Vaccinium oxycoccus*) occurs in these mires with some scarce mosses, herbs and sedges. In the wetter west, in Kirkcudbrightshire, the least disturbed area of peatland in southern Scotland occurs. This is known as the Silver Flowe and is one

The mixed sand and shingle shores of the Mull of Galloway hold colonies of oyster plant (Mertensia maritima), but the strongest populations occur along the Banffshire coast and on Orkney.

of the most important mire systems in Scotland. It also represents the most southerly development of this type of oceanic blanket mire which is so widespread in north-west Scotland. The Silver Flowe contains superb examples of the characteristic hummock-hollow bog surface; some areas approach the structure of raised mires.

Raised mires form on level flood plains of rivers especially over deposits of alluvium. With accumulation of peat the surface of the mire becomes dependent on rainfall rather than groundwater for its maintenance and growth. Raised mires characteristically form a low dome bounded by a stream or lagg. Oceanic raised mires are a feature of western Britain and in southern Scotland Kirkconnel Flow in Kirkcudbrightshire is a good example. A characteristic feature of southern Scottish and English raised mires is the widespread occurence of cranberry (*Vaccinium oxycoccus*) and bog rosemary (*Andromeda polifolia*).

Other features of botanical interest in southern Scotland include the Wigtownshire coast around the Mull of Galloway where northern cliff communities containing *Ligusticum scoticum* meet southern communities containing *Crithmum maritimum*. Several Scottish plants such as purple oxytropis (*Oxytropis halleri*) occur there at their southern limit, whilst the small restharrow (*Ononis reclinata*) reaches its northern limit. There are also some good populations of oyster plant (*Mertensia maritima*). On the sedimentary rocks at Southwick and Rockcliffe, Kirkcudbrightshire, the populations of the sticky catchfly (*Lychnis viscaria*) are a south-westerly Scottish outpost for this plant.

Midland Valley

Ben Lawers is the highest mountain in the Breadalbane range. It provides the greatest range of alpine habitats for plants on soft calcareous mica-schist rocks.

The Midland Valley lies between the Southern Upland Fault and the Highland Boundary Fault which runs north-eastwards from the Isle of Arran in the Clyde, through the south end of Loch Lomond, to Stonehaven on the east coast. The rocks are of mainly Devonian and Carboniferous age but the oldest exposures are Ordovician and Silurian sandstone. Despite its name, the Midland Valley (a rift valley) has plenty of upland areas which include the Pentland Hills, Carrick Hills, Renfrew Hills, Campsie Fells, Ochil Hills and Sidlaw Hills. The Ochils, to the north-east of Stirling, are composed of igneous andesitic and basaltic lavas. The soil is mildly acid, which favours *Lychnis viscaria*; this area has the greatest concentration of this plant in Britain. Further populations on outcrops of similar rocks in Glen Farg, Edinburgh (and formerly in other localities in the Lothians) make this plant a speciality of this geological area. Around the centre of the region spoil heaps or 'bings', legacies from the coal-mining industry, have provided a range of sites for plant colonization in an industrial heartland. Plants adapted to the unusual soils of these sites are able to exploit the conditions with some freedom from competition. One speciality, Young's helleborine (*Epipactis youngiana*), which is currently considered to be a newly evolved endemic species, has been recorded from four of these spoil heaps. In the north-east of the region, the sheltered base-rich coastal cliffs and dunes at St. Cyrus provide a warm site where a number of plants such as clustered bellflower (*Campanula glomerata*) reach their northern limit in Britain.

The Grampian Highlands

The Grampian Highlands are between the Highland Boundary Fault and the Great Glen Fault which traverses mainland Scotland from Inverness, through Loch Ness and Glen More to Fort William. There is a great contrast between the relatively continental Eastern Highlands and the oceanic west. The bulk of the rocks of the Grampians are schists of Dalradian (mainly Cambrian) and Moinean (Pre-Cambrian) age. Botanically, the extensive area of soft calcareous schists make the Grampian Highlands particularly rich, whilst the intrusions of acidic rocks of the Cairngorm massif form a distinct contrast. The long and indented coastline of the west has some of the finest oceanic woodland in Europe. The canopy of these humid deciduous woods is composed mainly of oak, ash, birch and hazel. They are often dissected by fast running streams with waterfalls and cascades. The rocky floors of the woods frequently make access dangerous with deep crevices between boulders obscured by a luxuriant growth of mosses, liverworts and ferns. Whilst there is an abundance and variety of 'lower plants', the flowering plants are rather less remarkable except for the spring shows of wood anemone (*Anemone nemorosa*) and bluebell (*Hyacinthoides non-scripta*). A small coastal area of special interest to the north of the Mull of Kintyre is a small outcrop of lime-bearing rock with a remarkable collection of alpines including colonies of purple and yellow saxifrage (*Saxifraga oppositifolia* and *S. aizoides* respectively). Most interesting is a peculiar form of yellow oxytropis (*Oxytropis campestris*) with purple-tinged flowers.

One of the most important botanical areas of Scotland is the calcareous mica-schist mountains of the Breadalbane range which extend eastwards from Beinn Laoigh in the west to Ben Lawers in the east. The extent of this range is sufficient for differences in climate to play a part in the distribution of certain species, besides variation in altitude. In addition to the many rare species which are sporadic in their occurrence, individual mountains often differ in the abundance of particular species. The soft, easily weathered calcareous rock creates a fertile soil and rockfalls, slumps and slides ensure renewal of niches for plants to colonize. Not all the crags are calcareous but the bright cushions of *Saxifraga oppositifolia*, moss campion (*Silene acaulis*) and *Saxifraga aizoides* are good indicators of botanically interesting ground. The higher cliffs are usually the richer areas and because of pervasive grazing pressure, places inaccessible to sheep and deer have the greatest concentrations of plants. Clefts, ledges and crevices of various aspects provide an endless variety of habitats. On Ben Lawers and the hills in the immediate neighbourhood a number of special plants are found including alpine forget-me-not (*Myosotis alpestris*), snow gentian (*Gentiana nivalis*) and drooping saxifrage (*Saxifraga cernua*). These are very rare but not absent outside the Ben Lawers area. However, Ben Lawers, the highest in the Breadalbane range at 1214m, offers a greater range of climate and therefore more varied conditions for plants. Ben Laoigh is particularly rich in alpine bartsia (*Bartsia alpina*) and mountain bladder-fern (*Cystopteris montana*) which probably enjoy the more oceanic climate of the westerly hills, whilst the less calcareous rocks hold the largest population of alpine woodsia (*Woodsia alpina*) in Britain.

To the east of the Breadalbane mountains there are further outcrops of calcareous Dalradian rocks at the head of Caenlochan Glen, Glen Clova and Glen Doll. Most of the Breadalbane specialities grow here, some not in such quantity but the area has its own treasures. On a broad ledge near Glen Doll the best relic of montane willow scrub survives which consists of a mixture of the rare woolly willow (*Salix lanata*) and downy willow (*Salix lapponum*). In the flushed grassland below, a colony of the even rarer close-headed Alpine-sedge (*Carex norvegica*) makes the area especially exciting. In Caenlochan Glen itself, on more acidic rocks the alpine blue sow-thistle (*Cicerbita alpina*) grows on a north-facing ledge, out of reach of most sheep and deer.

On the high plateau to the north-west of Glen Clova a small high-level outcrop of serpentinite rock holds by far the largest of the two British colonies of alpine catchfly (*Lychnis alpina*). Its sparse associates on this open fellfield include a very rare form of the common mouse-ear (*Cerastium fontanum* subsp. *scoticum*) and also beautifully domed cushions of cyphel (*Minuartia sedoides*). All these species belong to the pink family (Caryophyllaceae) which seems to have a predilection for this rock.

The glacial basin of Rannoch Moor, bounded on three sides by mountains, is one of the most desolate yet atmospheric places in Scotland, an undulating blanket mire with pools in the depressions. It is the last surviving home for what has become known as the Rannoch rush (*Scheuchzeria palustris*) which formerly grew in a number of English counties but is now extinct. The plant requires constantly wet conditions and is particularly susceptible to drainage.

Alpine bearberry (Arctostaphylos alpinus) is particularly tolerant of wind, and avoids competition by growing in the most exposed positions on the mountains of the north-west highlands.

10

In Perthshire and Angus some good examples of northern calcareous ash, oak and elm woodlands survive on the sides of precipitous ravines cut through limestone, mica-schists or base-enriched acidic rock. They are often named 'dens' on maps (Den of Airlie for example) and contain a rich assortment of flowering plants, ferns and bryophytes. Their inaccessibility protects them from grazing and the unstable soil, much like that on the calcareous mountains, continually provides new ground for colonization. This type of woodland reaches its geographical limit in central Scotland and some typical trees and shrubs of similar woodland in England, such as field maple (*Acer campestre*) and

small-leaved lime (*Tilia cordata*) are absent. These absences, and several northern plants such as marsh hawk's-beard (*Crepis paludosa*), mountain melic (*Melica nutans*), stone bramble (*Rubus saxatilis*) and wood crane's-bill (*Geranium sylvaticum*), impart a northern feel to these woods. In just a few localities small colonies of the extremely rare whorled Solomon's-seal (*Polygonatum verticillatum*) survive, but the plant is so difficult to spot at a distance that it may grow undetected in other ravines.

The outcrops of Durness limestone at both Inchnadamph and on the north coast are marked out by the displays of the lime-loving mountain avens (Dryas octopetala).

The massive area of the Cairngorm Mountains provides a great contrast to the Breadalbane range. The prevailing rock type is acidic granite, so the calciphile arctic-alpines characteristic of the Breadalbanes are largely replaced by calcifuge or lime-hating species. The lower slopes and valley floors are mainly covered with blanket mire and heather moorland in the drier areas. In these montane heaths so called 'peat alpines' are often seen in local abundance. One of the most attractive, dwarf cornel (*Cornus suecicus*) is characteristic of the central and western Highlands and in the Cairngorms may be accompanied by cloudberry (*Rubus chamaemorus*), chickweed wintergreen (*Trientalis europaea*), bog bearberry (*Vaccinium uliginosum*) and interrupted clubmoss (*Lycopodium annotinum*). With increasing altitude the heather becomes wind-pruned and assumes a more prostrate form and in exposed open ground the trailing azalea (*Loiseleuria procumbens*) often occurs in abundance. On wetter slopes, often on the sides of corries, where drainage water reaches the surface, soligenous mires form. These mires stand out as conspicuously bright patches of vegetation dominated by sedges, sphagnum and other mosses. They are very different from those on Ben Lawers where they are dominated by a suite of mosses that prefer more basic conditions. The hard rock walls of the corries appear bare in comparison with those of softer calcareous rocks but have their specialities such as brook saxifrage (*Saxifraga rivularis*), alpine speedwell (*Veronica alpina*), starwort mouse-ear (*Cerastium cerastoides*) and arctic mouse-ear (*Cerastium arcticum*). Indeed, the two similar mouse-ears *Cerastium arcticum* and *Cerastium alpinum* indicate differing conditions. *C. alpinum* is confined to soft calcareous rocks and is abundant in the Breadalbane mountains, whilst *C. arcticum* occurs on harder basic and acid rocks and has its headquarters in the Cairngorms.

11

On the high Cairngorm plateau the prostrate *Calluna* grades into extensive heaths of woolly hair-moss (*Racomitrium lanuginosum*) and three-leaved rush (*Juncus trifidus*). The vegetation of areas associated with snow-beds is extremely important for conservation and it is particularly rich in bryophyte-dominated communities. In only a few places in the Cairngorms does the rock type allow a flora similar to that of Ben Lawers to develop. This occurs notably in Coire Garbhlach on the western side of the Cairngorms where calcareous schist of Moinean (Pre-Cambrian) age outcrops. Another hill, Creag an Dail Bheag, on the south-east fringe of the range is also famous for its calcareous montane grassland, flushes and cliffs.

At the northern foot of the Cairngorms the forests of Rothiemurchus, Abernethy, Inschriach, Invereshie and Glen Feshie constitute the largest remaining tract of native pinewood in Scotland. They are not rich in flowering plants and are dominated mainly by dwarf ericaceous shrubs and bulky mosses, but these eastern Highland pinewoods have a small collection of beautiful but scarce plants which heighten the excitement of those botanizing in these atmospheric places: common wintergreen (*Pyrola minor*), serrated wintergreen (*Orthilia secunda*), creeping lady's-tresses (*Goodyera repens*), twinflower (*Linnaea borealis*) and, the rarest, one-flowered wintergreen (*Moneses uniflora*). Some old pine plantations also have fine colonies of these pinewood herbs. The natural altitudinal limits of the pinewood have been almost completely destroyed by grazing and fire, but on a single spur of the western Cairngorms, Creag Fhiaclach, a zone of dwarfed, gnarled pines marks the edge of the only remaining example of the natural treeline.

An indicator of moist, calcareous conditions, the yellow saxifrage (Saxifraga aizoides) occurs in abundance on limestone but often picks out areas of local flushing among more acid rocks.

The coastline between Nairn and Banff is varied and the geology around Banff and Buckie is especially complicated. Here, the Dalradian schist, with local areas of limestone and serpentinite, supports an interesting vegetation with several areas where calcicolous plants occur. Among them, purple saxifrage (*Saxifraga oppositifolia*) with mossy saxifrage (*Saxifraga hypnoides*) may be seen at their lowest eastern altitude. The shingle beaches of this stretch of coastline are important for the northern oyster plant (*Mertensia maritima*).

The largest area of fen in northern Britain is at the Insh marshes in the valley of the River Spey between Kincraig and Kingussie. The term 'fen' is used to describe peatland which occupies broad flood-plains. Vegetation is neutral to calcareous dominated by grasses, sedges and tall herbs with transitions to open water. The Insh fens have a great range of

plant communities varying with degree of wetness and nutrient status. The sedge flora, which includes the rare string sedge (*Carex chordorrhiza*), is of particular interest and the abundance of the northern water sedge (*Carex aquatilis*) and the least waterlily (*Nuphar pumila*) which grows in open pools, give a distinctly Scottish character to these fens.

Purple oxytropis (Oxytropis halleri) has a coastal montane distribution in Scotland growing on mountains, and sea cliffs especially along the north coast.

Northern Highlands

The area north of the Great Glen is another varied region and possesses many distinctive features of topography and vegetation. The geology is complex but except for the Devonian sandstone of much of Caithness, the rocks are mainly Pre-Cambrian (over six hundred million years old). The oldest rocks are Lewisian and predominantly acid, but they vary greatly and many local outcrops of more calcareous rocks occur. There is a narrow band of Durness Limestone of Cambrian/Ordovician age which stretches from Loch Kishorn north-west to Durness; this dramatically influences the vegetation where it outcrops. Younger rocks include a limited area of Tertiary basalt north of the Sound of Mull but this is much more extensive on the islands of Skye and Mull.

On the west coast the lichen- and bryophyte-rich oceanic woods continue northwards to Sutherland and become richer in some groups of northern lichens. The curious pipewort (*Eriocaulon aquaticum*) grows in a few lochans on the Ardnamurchan peninsula, and in the vicinity of Loch Shiel the Irish lady's-tresses (*Spiranthes romanzoffiana*) occurs in peaty grassland dominated by purple moor-grass (*Molinia caerulea*). These plants occur nowhere else in Europe apart from localities in the Hebrides and Ireland and have their main distributions in North America. The speciality of the otherwise botanically dull acid mountains around the head of Loch Shiel is the arctic diapensia (*Diapensia lapponica*). The mountain ridge on which this plant grows is composed of a hard mica-schist and supports many common and acid-tolerant montane plants, but the discovery of *Diapensia* in 1951 put the hill firmly on the botanical map.

Northwards, large tracts of low ground and lower hill slopes are dominated by *Molinia caerulea*, *Erica tetralix*, sparse *Calluna* and bog mosses. In areas of slightly higher nutrient status *Saxifraga aizoides* and broad-leaved cottongrass (*Eriophorum latifolium*) occur and, more locally, the delicate pale butterwort (*Pinguicula lusitanica*) give a western character to the vegetation. *Arctostaphylos alpina* is characteristic of the wind-blasted ridges of the northern Highlands where it accompanies *Loiseleuria procumbens*, bearberry (*Arctostaphylos uva-ursi*) and crowberry (*Empetrum* spp.).

The main area of western pinewood is around Beinn Eighe in Ross-shire; it almost reaches its western European limit at Shieldaig. These pinewoods have more oceanic mosses in the ground vegetation and the more strictly eastern pinewood herbs, *Linnaea borealis* and *Moneses uniflora*, are largely absent.

The only mountains that produce a calcicolous flora approaching the richness of the Breadalbane and Caenlochan mountains are those composed of Moine schists and gneiss in the Inverlael and Freevater Forests south-east of Loch Broom. Though these rocks are not generally as base-rich as the soft Dalradian schists of Ben Lawers, many of the special plants occur in the north-facing corries below the summits of Beinn Dearg and Seana Bhraigh. Again, a good indicator of the slightly less calcareous conditions is the common occurrence of *Cerastium arcticum* rather than C. *alpinum* – a situation reversed in the Breadalbane mountains. On the more acid, exposed summit plateaux in this area the very rare *Artemisia norvegica* grows, first discovered on a spur of one of the Torridonian Sandstone mountains near Elphin, north of Ullapool.

A little further north, at Inchnadamph, the widest exposure of the Durness Limestone occurs. Its open grasslands and sheltered cliffs possess rarities such as Norwegian sandwort (*Arenaria norvegica* subsp. *norvegica*) and dark red helleborine (*Epipactis atrorubens*), but it is the striking show of *Dryas octopetala* and *Saxifraga aizoides* that contrasts most with the surrounding moorland.

The two whitebeams (Sorbus arranensis and S. pseudofennica) are endemic to the island of Arran and have resulted from hybridization between rowan (S. aucuparia) and rock whitebeam (S. rupicola).

At the northern end of the limestone outcrop, around Durness itself, is the largest area of *Dryas* heath in Britain, almost rivalling that in County Clare in western Ireland. The climate of the north Scottish coast is particularly harsh with no shelter from the bitterly cold northerly winds. Arctic-alpines descend to low altitudes on the headlands and *Silene acaulis, Saxifraga oppositifolia* and *Oxytropis halleri* join *Dryas* on the base-rich rocks, whilst on the acid moorland behind Cape Wrath the alpine dwarf shrub heath with *Loiseleuria procumbens* and *Arctostaphylos alpina* descends to below 300m. The Scottish primrose (*Primula scotica*) nestles in the grazed turf immediately behind the steep coastal slopes in the damp sea-sprayed maritime heath and grassland. Even scarcer than the beautiful Scottish primrose are a number of coastal eyebrights which sometimes grow near *Primula scotica*. Some are very attractive but because they are difficult to identify they are ignored by many botanists. Scots lovage (*Ligusticum scoticum*) occurs, amongst the taller cliff vegetation along the north coast, tucked away in crevices that escape the brunt of the most violent storms.

The characteristic rock pinnacles of the Storr on the island of Skye rise above the damp basalt screes where the rare Iceland purslane (Koenigia islandica) grows.

Away from the coast and extending right across north-east Sutherland and part of Caithness the Flow Country forms the largest expanse of blanket bog in Scotland. With its varied and intricate patterns of pools, it is not renowned for its diversity of flowering plants and is dominated by dwarf shrubs, sedges and sphagnum mosses. However, it is internationally important for its physical structure and patterns of vegetation. Over such a vast area changes in vegetational structure from east to west are noticeable. Oceanic species such as bog myrtle (*Myrica gale*) and bell heather (*Erica cinerea*) are more abundant in the west, and *Molinia caerulea* is widespread, whereas in the east it is confined to wet flushes.

Around the Dornoch Firth in east Sutherland, outcrops of dry granitic rocks intruded into the older Moinean rocks have a more interesting flora than might be expected because minerals in the granite alleviate the low nutrient status. The extremely rare rock cinquefoil (*Potentilla rupestris*) occurs in two places and pyramidal bugle (*Ajuga pyramidalis*) may be found below the cliffs. On other rocks in Sutherland, especially in the south-west where base status is slightly raised, *A. pyramidalis* is characteristic of open heathy grassland in the vicinity of birch woodland.

The Islands

Islands generally have fewer plant species than adjoining land-masses and their relative isolation may give rise to local races which may form new species. In the south, the Isle of Arran in the Firth of Clyde is notable for its two endemic whitebeams, Arran whitebeam (*Sorbus arranenis*) and Arran service-tree (*Sorbus pseudofennica*). These occur in rocky glens cut into the massive tertiary granite intrusion which forms most of the northern half of the island. It remains a mystery why these species, which are hybrids between rock whitebeam (*Sorbus rupicola*) and rowan (*Sorbus aucuparia*), should only occur in Arran as the two parent species can be found together elsewhere.

Pipewort (Eriocaulon aquaticum) has its main centre of distribution in North America, but within Europe it is confined to lochans in Northern Ireland, the Inner Hebrides and on the Ardnamurchan Peninsula.

One of the most interesting features of the Inner Hebrides is the outcrops of basic volcanic rocks of Tertiary age (beginning 65 million years ago). These rocks form part of the Tertiary Igneous Province, a once vast volcanic region active during this period. They are mainly restricted in the British Isles to north-east Ireland, Ardnamurchan and Morvern, Arran and the Inner Hebrides, but the same system embraces Faeröe and outcrops again in Iceland and east and west Greenland. On the Ardmeanach Peninsula on Mull and on the Trotternish ridge on Skye, the basalt weathers to form open gravel pans and fine screes where the arctic annual *Koenigia islandica* reaches its southern limit in the northern hemisphere. The basaltic north-facing cliffs of Trotternish are unrivalled scenically with slipped blocks, rock pinnacles and block screes. The more stable cliffs in more calcareous areas form the richest locality for mountain plants in the Hebrides.

Grassland overlying the basalt at lower altitudes and on sea cliffs provides mildly calcareous grassland rich in wild thyme (*Thymus polytrichus* subsp. *britannicus*), on which the scarce thyme broomrape (*Orobanche alba*) is parasitic. This southern European species reaches its northern limit in the western Scottish Highlands and Islands. The abundance of basalt rock and hence the abundance of its host may account to some extent for the concentration of localities in this area.

The lochans around Sligachan on Skye constitute the main area for *Eriocaulon aquaticum* although it is also found on Coll and the Ardnamurchan Peninsula. Another member of the American Element of the British flora, *Spiranthes romanzoffiana*, is not present on Skye but again occurs on Coll, and also on Colonsay, Islay, Mull and in the Outer Hebrides. The Outer Hebrides compared with most of the other Scottish Islands, except perhaps for the Old Red Sandstone of Orkney, are geologically uniform.

They are largely composed of acid Lewisian gneiss. The peatlands of Lewis in particular include large areas of blanket mire with pool systems as well as some fine examples of valley and basin mires set amongst the low undulating outcrops of rocks. On Benbecula and Barra *Spiranthes romanzoffiana* occurs around peaty lochans and old lazybeds. The machair grassland formed along the western coastal strip is the major botanical interest of the Outer Hebrides. The main attraction is the sheer abundance of flowers rather than rare species. The Hebridean race of the common spotted-orchid, the Hebridean spotted-orchid (*Dactylorhiza fuchsii* subsp. *hebridensis*), often occurs in abundance, together with the deep red form of the early marsh-orchid (*Dactylorhiza incarnata* subsp. *coccinea*). One speciality of the machair of North Uist is the attractive variant of the broad-leaved marsh-orchid (*Dactylorhiza majalis* subsp. *scotica*) known from nowhere else.

The outcropping rocks of the Northern Islands of Orkney and Shetland are very different from each other geologically. The former is almost entirely composed of sedimentary rocks of Old Red Sandstone age (Devonian) whilst Shetland contrasts in its complexity. The rocks of Shetland have been very much affected by episodes of vulcanicity and mountain building which have altered them by heat and pressure whilst igneous intrusions have added further diversity to the bedrocks. The Great Glen Fault runs east of Orkney and passes through the Shetland Islands. However, the climate of the two island groups is similar and on Orkney the severity of the exposure is indicated by the abundance of *Arctostaphylos alpina* with other species such as *Loiseleuria procumbens* in the montane dwarf shrub heath which descends to possibly its lowest level in Britain behind the cliffs near the Old Man of Hoy. Maritime grassland and heath in several areas of Orkney are famous for their large populations of *Primula scotica*, whilst the shingle beaches carry the most extensive colonies of *Mertensia maritima* in Britain.

The vegetation of Shetland is much more diverse. The isolation of this group of islands has resulted in many plants being recognized as local varieties, subspecies or even species. Among this diversity the Keen of Hamar, a low hill on the north-east coast of Unst, is particularly fascinating. Here it seems that several extremes of climate and soil have combined to create very exacting conditions for plant life. The serpentinite bedrock forms part of a belt which runs north-east to south-west through Unst including a large area of Fetlar. Exposures of serpentinite on the Keen of Hamar differ from others in Shetland by the large areas of open sparsely-colonized rock debris and rarities such as the endemic Shetland mouse-ear (*Cerastium nigrescens*) and *Arenaria norvegica* subsp. *norvegica* occur in local abundance. The unusual chemistry of serpentinite soils, typically with high levels of heavy metals and very low availability of phosphate, demand adaptation by plants. The Keen of Hamar soil does not appear to be different chemically from the other Shetland serpentinite soils but the physical structure of the rock appears to be unusual. Weathering shatters the rock on the Keen of Hamar in a characteristic way and forms a shallow free-draining soil. Plants have often been observed wilting in summer, even in the oceanic climate of Shetland with 295 rain-days (days with >0.2 mm rainfall) per year. Further, a number of plants on the Keen of Hamar show adaptations to very dry conditions. *Cerastium nigrescens* has fleshy leaves and a dense covering of glandular hairs whilst the local form of sea plantain (*Plantago maritima*) is extremely woolly. From these observations, ecologists consider that it is the excessively dry conditions during the summer that is mainly responsible for the lack of colonization of the open areas of rock debris.

HUMAN INFLUENCES AND CONSERVATION

There is little evidence of the influence of man on the vegetation of Scotland before forest clearance for grazing and early cultivation began to make its mark about 3000 years ago. Since then vegetation development and the distribution of plants and animals has become progressively influenced by human activities. The changes brought about by these activities range from subtle shifts in the abundance of certain species to total habitat destruction.

There is only one recorded locality, in the western Cairngorms, where native forest reaches its natural altitudinal limit. Elsewhere the treeline has been lowered by grazing, burning and cutting. Above the natural forest limit, especially on base-rich mountains, willows potentially form a characteristic vegetation zone. The alpine willows are particularly sensitive to grazing and this montane scrub has been reduced to a few fragments in Scotland, confined to ledges inaccessible to grazing animals. Montane tall-herb communities have been similarly affected. *Cicerbita alpina*, a common plant of montane tall-herb vegetation in Norway, but very rare in the Scottish Highlands, is probably a relic of a once more intact cover of sub-alpine woodland.

Scotland's extensive heathlands have been largely formed through forest clearance, although in some areas, the onset of a cooler, wetter climate in Sub-Atlantic times favoured ericaceous sub-shrubs and depressed tree growth. Here man may have merely hastened a natural process. Unmanaged heathland has a diverse age structure. The cycle of regeneration, growth and death of heather plants creates a mosaic of gaps which allows other kinds of plants to invade. Burning heather to provide food for grouse and sheep acts as a form of pruning, artificially maintaining heather in a vigorous condition. Frequent burning on a large scale destroys the natural mosaic of age classes and results in a loss of diversity. Fire also damages the natural transition between heathland and mires by drying the edges of the latter and creating more favourable conditions for heather growth, thus extending the heathland. Constant heavy grazing weakens heather and allows grasses to dominate and transforms heathland to acid grassland.

Neutral or calcareous grasslands were mainly formed when deciduous woodland over more basic rocks was cleared and tree regeneration was prevented by grazing animals. These grasslands require grazing or cutting to prevent them from reverting to woodland. Conversely, heavy grazing favours only the most resistant species and tends to reduce botanical diversity. Some plants are able to adapt to these conditions, within limits, but less-adaptable species decline. Modern grassland management, which aims to increase productivity for pasture or silage, involves various activities including drainage, fertiliser application and in some cases re-seeding with cultivated varieties of vigorous nutrient-rich grasses. These reduce the diversity of grassland communities and favour very few grasses and herbs.

Many of the plants selected here are rare or scarce. Natural rarity is an expression of a plant's response to ecological and geographical factors. All plants have a certain range of tolerance to environmental conditions, whilst the biology of a species determines its population size. The rarest plants have narrow geographic ranges, exacting habitat requirements and a tendency to form small populations. *Polygonatum verticillatum* in its few wooded ravines in Perthshire is an example of this in Scotland. Many rare plants may not be particularly threatened with extinction because their habitats are remote or relatively free from human influences.

The Hebridean spotted orchid (Dactylorhiza fuchsii subsp. hebridensis) is one of the specialities of the calcareous machair grassland of the Outer Hebrides.

The most vulnerable are those that grow in habitats much affected by management. For instance, over 90% of the total British population of *Moneses uniflora* occurs in pine forests managed to some extent for timber production. Plantation forests may have created new habitats for this rare plant in the face of drastic reduction of native pinewood, but the well-being of these colonies now depends on sympathetic forest management which avoids clear-felling in the vicinity of the plants.

The conservation of individual species demands an understanding of their biology, ecology and habitat. For long-term survival it is often not enough to maintain the appropriate habitat type, but to keep it in a condition which allows particular species to regenerate and compete successfully with associated vegetation.

Gentiana nivalis grows in montane calcareous grassland on Ben Lawers. The grassland must be grazed within certain limits to enable the gentian to survive. If grazing is too light, vigorous grasses out-compete the plant, but if it is too heavy it prevents the species from setting seed. Similarly, grazing levels affect the performance of *Primula scotica* in northern coastal heath and grassland. The primula appears to grow most vigorously in coastal grassland where grazing keeps the dominant grasses in check and prevents the accumulation of dead leaf litter which hinders seed germination. Too high a stocking density renders the grassland vulnerable to erosion in severe storms. Grazing levels on maritime heathland seem to be less critical to maintain conditions suitable for *P. scotica*. Overgrazing converts the heathland to more fragile grassland which may be beneficial, but it is probably wiser to maintain a mosaic of the two habitat types. Achieving an optimum grazing regime requires considerable time and effort and invariably benefits from the knowledge of local graziers.

Site-based conservation can be achieved through the various designations of land protection, principally Sites of Special Scientific Interest and National Nature Reserves, some of which are owned by statutory and non-governmental organisations. On privately owned land conservation may be secured through management agreements with landowners or occupiers. For example, a management agreement might provide for the continuation of traditional farming practices to maintain plant communities of conservation importance. The effects of management, experimental or otherwise, must be monitored to determine whether the desired result is being achieved or whether it needs to be modified.

All wild plants are protected in the United Kingdom by the Wildlife and Countryside Act (1981) which prohibits uprooting without a landowner's permission. Schedule 8 of this Act prevents the collection of any part of our most threatened species. The list of Schedule 8 plants is reviewed every five years to take into account any changes in the status of a species.

Besides conserving plants in the wild, heightened pressures on the environment makes it increasingly necessary to provide safe sites for our most threatened species. Many botanic gardens have provided this facility for some time but their role in *ex situ* conservation has greatly accelerated within the last 20 years. Greater knowledge of, and concern for, the conservation of genetic variation within species has increased the need to store large quantities of plant material. One of the main methods is the use of seed banks and, more recently, spore banks. A spore bank has been used to store spores and regenerate plants of alpine woodsia (*Woodsia alpina*), one of our most threatened ferns. At very low temperatures and humidity, seeds of many species can be stored for long periods with little loss of viability, but plants are regenerated at intervals to ensure fresh supplies of seed for the long term. Botanic gardens play a vital role in conservation through their collections, their research into the propagation and cultivation of plants and by displaying the diversity of the plant kingdom.

Conservation depends heavily upon public understanding and appreciation of our native wildlife. The greater the familiarity with our flora, the greater is the concern for its well being. To instill sustained public interest in the natural world, environmental education is of paramount importance. We hope that the selection of plants presented in this book will help to increase the knowledge and familiarity of a range of our less well-known plants whilst setting some of the better known Scottish plants in a wider context.

The distribution maps on the following pages include both historic and current records for each plant species. Some of Scotland's more threatened plants, *Moneses uniflora* for example, are presently known from fewer vice-counties than those shown.

The Plants

Ajuga pyramidalis L.
PYRAMIDAL BUGLE

Of the three native species of ajuga in Britain, bugle (*Ajuga reptans*) is widespread, whilst the ground pine (*Ajuga chamaepitys*) and pyramidal bugle are geographically restricted. *A. chamaepitys* is a plant of open chalk downland in the south of England and *A. pyramidalis* is almost confined to Scotland. Indeed the two species are rather like northern and southern counterparts.

Ajuga pyramidalis was first reported as a Scottish plant in John Lightfoot's *Flora Scotica* of 1777. As this was apparently second-hand information from Rev. John Burgess, parish minister of Kirkmichael, Perthshire, without mention of a specific locality or specimen as proof, the record was doubted in other botanical works of the period, such as James Sowerby's *English Botany*. The first record attached to a definite locality was that by Dr D. Hope from Ben Nevis in the second edition of William Hudson's *Flora Anglica* of 1778. Another very early record is from Tor Achilty, which was described then as (and still is) a beautiful wooded hill overlooking Loch Achilty in East Ross-shire. The plant is still known from this locality.

Ajuga pyramidalis, East Ross, 21 v 1995.

Ajuga pyramidalis is most abundant in south-west Sutherland which can be considered its centre of distribution within the British Isles, but the plant extends to the Western Isles and reaches counties Galway and Clare in Ireland. Outlying populations in the south occur in the Moffat Hills where it is extremely restricted and in one place is accompanied by the rare fern *Woodsia ilvensis*. Pyramidal bugle reaches its most southerly British limit on Ill Bell in Westmorland where it grows at an altitude of 650 metres. It was discovered there by John Backhouse in 1869.

In the Scottish Highlands *A. pyramidalis* tends to grow close to woodland in open places where the ground is broken on warm south-facing slopes which catch the spring sunshine. The plant occurs on acid to slightly basic soil overlying a range of rocks such as Lewisan gneiss, Silurian shale, Old Red Sandstone and, in at least one place near Spinningdale in south-east Sutherland, on granite containing basic minerals.

In Ireland the plant occurs on carboniferous limestone, but whatever the soil the plant generally favours some freedom from competition. Its associated flora is generally that of dry acid hillsides including heather (*Calluna vulgaris*), bell heather (*Erica cinerea*), purple moor-grass (*Molinia caerulea*), tormentil (*Potentilla erecta*), velvet bent (*Agrostis canina*) and heath milkwort (*Polygala serpyllifolia*), but it is also accompanied by other plants that are more abundant in open vegetation such as slender St John's-wort (*Hypericum pulchrum*) and mountain everlasting (*Antennaria dioica*). Occasionally the scarce thyme broomrape (*Orobanche alba*) grows in the vicinity. *Ajuga pyramidalis* also occurs at the foot of the crags of one of the two Sutherland localities of rock cinquefoil (*Potentilla rupestris*).

Pyramidal bugle is mainly restricted to Europe and extends to the north of Norway and reaches Iceland as a rarity. Its south and eastern limits are central Spain and Turkey respectively.

A. pyramidalis is early flowering, commencing in May. The flowers, like many in the family Lamiaceae, are adapted for insect pollination with the lower lip of the flower forming a landing platform for winged insects. A feature of *Ajuga* is the extremely short upper lip of the flower which makes the lower lip even more obvious as a platform for potential pollinators such as bumble-bees and hover-flies. The short upper lip exposes the stamens and stigma although they are protected from rain by the leaf-like bracts above them. Nectar is produced at the base of the corolla tube and the stamens mature slightly before the stigma, thus promoting cross-fertilization. The style is initially positioned above the stamens but after dehiscence of the anthers it bends downward bringing the stigma into the entrance of the flower tube where it is brushed by the visiting insects. Ants, which are attracted by a glandular secretion on the lower lip, are not effective pollinators but have a role in seed dispersal. At the end of the fruit there is an oil body (elaiosome) which acts as food and seed production coincides with ant nest building and brood-rearing time. Ants have been recorded removing *Ajuga* seeds to their nests and then discarding them on the surface once the fatty tissue has been eaten. This benefits *Ajuga* seed as they require light for germination. Observations of ant seed dispersal have been made for *Ajuga reptans* but although *A. pyramidalis* has often been found growing in the vicinity of ants nests actual seed transport has not been seen. Unlike *Ajuga reptans*, *A. pyramidalis* does not possess long runners by which to spread vegetatively, so is more reliant on seed for dispersal. It may sometimes produce short shoots from buds in the axils of basal leaves which develop into young plants, but this habit is rather variable. Plants often die after flowering without producing these short shoots, totally depending on seed for continuity.

The sudden abundance of *A. pyramidalis* after fires has been observed and it seems that increased light reaching the ground stimulates the germination of buried seed. Other members of the *Ajuga* family behave in this way, notably the very rare downy woundwort (*Stachys germanica*) which appears, for example, after old hedges have been laid and the ground disturbed. The seed of *Ajuga* and *Stachys* remain viable in the soil for many years.

Arctostaphylos alpinus (L.) Sprengel

Mountain Bearberry

Although John Lightfoot has been credited with the discovery of mountain bearberry, it was James Robertson who first recorded it, in 1767, on Ben Horn in east Sutherland. He also found it on several other mountains in Sutherland and on Ben Nevis – the plant's most southern locality. Robertson extended the distribution of *A. alpinus* to Orkney, but not as far as Shetland, nor Skye. Lightfoot's 1772 record from the "misty Isle" stood alone for the next two centuries; no one could refind the plant on Ben-na-Greine, near Kyleakin, where he had seen it. Determined searches were made in 1986 and 1987 to no avail, though rediscovery of an alpine hawkweed, also unseen since Lightfoot's visit, and trailing azalea (*Loiseleuria procumbens*), a new record for Skye, were promising omens. Finally, the Kyleakin hills gave up their long-kept secret, and in 1990 *A. alpinus* was refound by Catriona Murray on Beinn Bhuide, a little to the south-east of its original locality. The reappearance of the plant was attributed to the removal of sheep from the lower hill. One wonders what other botanical treasures might be refound in certain 'lost sites' if freed from heavy grazing.

No botanist with any sense of self-preservation would linger long in *A. alpinus* territory, for the plant typically hugs the ground of wind-blasted ridges and bare rock-strewn shoulders below the summits of high mountains in the cold, wet north-west Highlands. Only on the extreme north coast and on the northern isles does it descend much below 600m, the potential tree-line.

The total distribution of *A. alpinus* is circumpolar. It occurs widely in Norway and Arctic Europe, but not in Iceland, and on the continent only in in the Alps and the Pyrenees.

Arctostaphylos alpinus in flower, Sutherland, 27 v 1995.

Although both species of *Arctostaphylos* can be found together, the commoner bearberry (*A. uva-ursi*) generally occupies lower ground and more sheltered parts of slopes, whereas *A. alpinus*, true to its name, is charcteristic of higher ground in low-alpine vegetation. It is a consistent component of *Calluna*-reindeer moss heath, together with other arctic-alpines, notably trailing azalea (*Loiseleuria procumbens*), mountain crowberry (*Empetrum nigrum* subsp. *hermaphroditum*), stiff sedge (*Carex bigelowii*), fir and alpine clubmosses (*Huperzia selago* and *Diphasiastrum alpinum* respectively) and the lichen, Iceland moss (*Cetraria islandica*). The prostrate stems of *A. alpinus* often grow through a lichen-rich carpet and the dwarfed heather, forming a woody mat no more than 10cm high

Fruit of Arctostaphylos alpinus, Sutherland, v 1982. (Photo: P Lusby).

which is broken into a mosaic by patches of stones. The species also occurs, much less typically, in high level blanket bog with a quite different set of associates, including hare's-tail cottongrass (*Eriophorum vaginatum*), cowberry (*Vaccinium vitis-idaea*), dwarf birch (*Betula nana*) and cloudberry (*Rubus chamaemorus*).

A. alpinus is instantly recognizable and distinguished from other heaths by its deeply impressed leaf veins which give the leaves a wrinkled appearance. Beneath, the leaves have a silvery lustre which is due to the two lowest layers of cells being separated from the rest. The scale-leaves, which surround the flower buds throughout the winter, have the same sandwich structure, but with an even thicker filling of air. This may act as insulation for the buds and young shoots against extreme cold and water loss. The flower buds of *A. uva-ursi* are unprotected by scale-leaves, and its foliage leaves remain green for two to three years.

The autumn colour of *A. alpinus* is one of the plant's most endearing features, turning the grey rocks over which it spreads a rich blood-red. In its wind-exposed habitats, *A. alpinus* does not receive any protection from snow and its deciduous habit is a means of reducing water loss and frost damage to a minimum in winter.

Autumn foliage of Arctostaphylos alpinus, Sutherland, 18 x 1995.

Mountain bearberry is an early-flowering heath, opening its buds in May, or as late as June on Ben Nevis. The species fruits rather sparingly in Scotland which may reflect a general scarcity of pollinating bumble-bees at this time of year. Several features of the flowers and reports of abundant fruit production in the Arctic, suggest that self-pollination occurs more regularly here, where the level of self-fertility may also be higher.

Any fruits which do form are fully ripe by July or August. The shiny black berries, the size of sloes, reputedly taste of blackcurrants and may be relished by sweet-toothed bears, whence both the plant's common and Latin names [*arktos* = Greek for bear, *staphyle* = a bunch of grapes]. There are records of the fruit being eaten by reindeer and foxes, but the most important seed dispersers in Scotland are probably grouse and ptarmigan; the Norwegian name of the plant is Rypebær, Grouseberry. These birds do not search for food by flying high and dipping to the ground as do smaller fruit-eaters such as thrushes. Rather, they settle and forage for fruits by walking amongst the heather, consuming berries in large quantities. It may be that a bright fruit colour (red) is not required to attract these birds. Nevertheless, the red foliage of *A. alpinus* may serve as a flag, advertising the presence of a meal to ground-foraging and high-flying species alike.

Artemisia norvegica Fries

NORWEGIAN MUGWORT

From the sloping windswept mountain spur in West Ross where *Artemisia norvegica* was first added to the British flora in 1950, one looks out on some of the finest Scottish mountain scenery with the shapely but massive bulks of Suilven, Stac Polly and Quinag rising majestically from the basement of Lewisian rocks.

Of the four species added to the Scottish mountain flora in the 1950s, *Artemisia norvegica* was perhaps the most exciting. Unlike *Koenigia islandica*, *Diapensia lapponica* and *Homogyne alpina*, which are decidedly more abundant in Scandinavia or Central Europe, *Artemisia norvegica* is extremely restricted in its global distribution. Outside Britain it is known only from central and southern Norway and the northern Ural Mountains. The Norwegian claim to the plant, upheld in the common name of Norwegian Mugwort, is understandable because it was from the central Norwegian mountains of the Dovre-Trollheimen area that the plant was first discovered in 1780 and was considered endemic to Norway until it was found at Chalmersdale in the Ural Mountains in 1848.

It was discovered in Scotland by Sir Christopher Cox and seven years later its range was extended from West Ross to the Ben Dearg-Seanna Bhraigh area of East Ross by H. Milne-Redhead. In 1971 it was discovered farther east on a ridge in the Freevater Forest by Dr Hugh Lang.

All Scottish sites for *Artemisia norvegica* are on extremely exposed mountain ridges composed of acid rocks at an altitude between 700 and 900m. At the classic West Ross locality, *Artemisia* grows in gravel derived from Torridonian sandstone, but in East Ross the rocks are mapped as moine schists, gneiss, granulites and quartzites. Although the plant occurs on acid rocks in Scotland and several of its sites in Norway, the Danish botanist Eric Hultén considered *A. norvegica* as "Kalkhold", meaning favourably influenced by lime although not confined to calcareous rocks. Although the soils on which it grows are generally acid and hence low in calcium and phosphorus, analysis of the West Ross soil has shown it not to be particularly base-poor and quite rich in potassium and magnesium. This is reflected in the group of plants associated with *Artemisia* which, although consisting mainly of an acid community of stiff sedge (*Carex bigelowii*) and woolly hair-moss (*Racomitrium lanuginosum*) with frequent alpine lady's mantle (*Alchemilla alpina*), the presence of plants such as moss campion (*Silene acaulis*), viviparous bistort (*Polygonum vivipara*), common wild thyme (*Thymus polytrichus* subsp. *britannicus*), cyphel (*Minuartia sedoides*), spiked wood-rush (*Luzula spicata*) and dwarf cudweed (*Gnaphalium supinum*), indicate less acid conditions. This suite of plants tends to occur where the effects of freezing and thawing are greatest and the regular disturbance of the soil helps to release plant nutrients.

The Scottish, Norwegian and Ural populations of *Artemisia norvegica* have been recognized as three distinct varieties: var. *scotica*, var. *norvegica* and var. *uralensis* respectively. The Ural variety is most clearly distinguished from Scottish and Norwegian plants by being glabrous throughout, lacking the silky-white hairs. Scottish plants are generally smaller in stature and also lack the distinct secondary leaf lobes of the Norwegian variety. In Scotland, flowering occurs in mid- to late-July with most plants producing a single head although multi-headed individuals occur occasionally.

Artemisia norvegica, West Ross, 9 viii 1995.

In Norway *Artemisia* is not quite so restricted to exposed mountain plateaux as in Scotland and occurs in rock crevices, on ledges and small terraces. However, the coldness and extreme exposure of most of its habitats make severe demands upon the plant. On some Norwegian mountains *Artemisia* is considered to be one of the most resistant plants to wind exposure. The very flat basal leaf rosette allows the plant to benefit from the influence of much higher temperatures at the soil surface compared with the surrounding air. In dry sunny conditions, ground temperatures in summer may rapidly rise and a maximum basal leaf temperature of 38°C has been recorded for *A. norvegica*. Although such high temperatures may not often be reached, the plant must be able to withstand them. The lethal temperature for leaves of *A. norvegica* has been found to be 47.9°C. In wetter habitats, where moisture at or near the surface has a cooling effect, plants are not subjected to such elevated temperatures. This is the case with *Koenigia islandica* which grows in similar, but constantly wetter habitats than *Artemisia* and hence has a lower lethal temperature. A further effect of dryness and exposure is the high potential water loss through the leaves. To reduce this, *Artemisia* has a leaf cuticle highly resistant to water passage. The silky-white hairs of Scottish and Norwegian plants further reduce water loss by creating an envelope of still air around the leaves.

In a few places in Norway *A. norvegica* has been recorded at lower altitudes and in less exposed situations where the plant may grow to a height of 50cm. This apparent flexibility in growth habit makes its extremely restricted distribution all the more puzzling. Poor seed dispersal has been suggested as the cause but has not been investigated. Dr S. M. Walters, in his and John Raven's classic book *Mountain Flowers,* suggests that *Artemisia norvegica,* like some other montane plants, is presently growing in environmental conditions that allow vegetative propagation but are unsuitable for seed production in normal seasons. Stiff sedge (*Carex bigelowii*) and several other species of plants in northern habitats have been found to survive purely vegetatively over thousands of years. It may well be that only very occasional seed-set by normal cross pollination is required to maintain these long-lived species in a healthy genetic condition.

Bartsia alpina L.

ALPINE BARTSIA

The hemi-parasitic alpine Bartsia was first found in Britain in 1668 near Orton in Westmorland (now Cumbria) by John Ray. He christened this previously undescribed species, *Euphrasia rubra Westmorelandica foliis brevibus obtusis* [Red Westmoreland eye-bright with short, blunt leaves]. More than a century later, in 1789, George Don claimed the first Scottish record from "rocks on the east side of Malghysdy" [= Meall Ghaordie] in the Breadalbane mountains.

The plant can still be found at both sites and also in England, in Middle and Upper Teesdale, and the north Yorkshire Pennines. The Breadalbane range of west Perthshire and Argyll is the plant's Scottish home and British stronghold. Although once known from the lower slopes of Ben Lawers, it has not been seen there for some years and may have disappeared due to heavy grazing. However, there have been more documented losses south of the border this century. In total, less than 30 populations remain.

Bartsia alpina, Argyll, 29 vi 1995.

Smith and Sowerby, in a volume of their *English Botany* in 1807, described the plant's preferred habitat as: "moist stony soil on the borders of alpine rills or little boggy spots in the interstices of rocky precipices". In Scotland *B. alpina* is very much a plant of ungrazed rock ledges which carry a rich, varied and often luxuriant mixture of dwarf shrubs, tall herbs, sedges and grasses amongst a carpet of cushion herbs and bryophytes. *B. alpina* thus has a long list of associates of diverse ecological affinities, including the northern-montane globeflower (*Trollius europaeus*), tall-herbs of fens and meadows such as wood crane's-bill (*Geranium sylvaticum*), the maritime-montane thrift (*Armeria maritima*) and sea plantain (*Plantago maritima*), heaths such as mountain avens (*Dryas octopetala*) and bilberry (*Vaccinium myrtillus*), and many other alpines, including purple saxifrage (*Saxifraga oppositifolia*), alpine mouse-ear (*Cerastium alpinum*) and moss campion (*Silene acaulis*). Apart from freedom from grazing and trampling, the vegetation is strongly determined by the calcareous nature of the bed rock. The limestone of Teesdale and the calcareous schist of the Breadalbane mountains make them classic sites for a rich alpine flora and *B. alpina* is one of several species common to both.

Outside Britain *B. alpina* occurs in Greenland, eastern Canada, Iceland, Faeröe and Scandinavia, eastwards to the Ural range. It is present throughout virtually all the mountain ranges of central Europe, a distribution which has earned it the common name, poly-mountain, latinized as *Polium montanum*. Like many arctic-alpines, *B. alpina* descends to a lower altitude towards its northern limit. Its preferred habitats are alpine meadows, lake shores and clearings in sub-alpine forests. In Arctic Norway it grows on bogs, damp cliffs, riverside flats and, exceptionally, below the shore line. In the lowlands of southern Scandinavia and central Europe it can be found as scattered relict populations in rich fens.

The genus *Bartsia* commemorates the German botanist Johannes Bartsch (1709–1738) who died prematurely from a tropical fever in Surinam. Carl Linnaeus had recommended him to a post there after turning it down himself on the belief that he would not survive the tropical climate. Linnaeus perhaps eased his guilt after hearing of his friend's death by immortalizing Bartsch's name in a plant dressed in mourning colours. On drying, the plant turns completely black, giving it a funereal starkness. *B. alpina* is known in Scandinavia as svarttopp [black-head].

The stiff little spikes of dark metallic-purple leaves and flowers recommend *B. alpina* for the rock garden or alpine trough. Much as its appearance may be valued, the plant's semi-parasitic habit has not favoured its widespread cultivation. Reginald Farrer, who sought out every sort of alpine plant, believed the only way to introduce it into a garden would be to "bring it down in a great sod from the mountain with all its companions included".

B. alpina utilizes a wide array of host plants, among them cowberry *(Vaccinium vitis-idaea)*, downy birch *(Betula pubescens)*, bog-rosemary *(Andromeda polifolia)*, alpine milk-vetch *(Astragalus alpinus)*, woolly willow *(Salix lanata)*, common butterwort *(Pinguicula vulgaris)* and several sedges and grasses. Plants of *B. alpina* have been found with no parasitic connections, but these are tiny with only a single flower. The main benefit derived from parasitism is nutritional, but *Bartsia* can also tap into its host's water supply. The leaves of *B. alpina* have a very low resistance to water loss, but it is able to afford a high rate of transpiration even in dry conditions as long as it is connected to the deeper roots of its host. Under severe moisture stress it is the host, not the parasite, that wilts first. Plant collectors know well that hemi-parasitic plants wilt almost as soon as they are cut.

However, a healthy supply of nutrients and water does not alone ensure reproductive success. *B. alpina* is dependent on insects for its pollination, of which bumble-bees are the main agents. A high degree of out-crossing is normally assured by the fact that cross-pollen grows faster than self-pollen and therefore is more likely to be first to reach and fertilize the ovules. Self-sired embryos are also generally aborted.

B. alpina can suffer heavy pre-dispersal seed losses due to predation by the larvae of a specialist micro-lepidopteran, and a fly which feeds on *Bartsia* and the related hemi-parasitic genera, *Pedicularis* and *Rhinanthus*. The larvae of both these predators can destroy a third or more of the seeds. Furthermore, apparently undamaged seeds in an attacked capsule have a much lowered germination capacity. After feeding, the larvae leave the capsules in late summer to pupate in the soil beneath their host plant. The violet bracts of *B. alpina* which contribute to pollinator attraction in pre-flowering may also attract predators. Plants with large flower spikes attract more pollinators but also suffer higher rates of predation than plants with small spikes.

Despite the seeds having wings, wind is not very important in their dispersal. However, moistened seed-wings stick efficiently to smooth surfaces such as birds' feet and in this way can be carried short or long distances. As long as seeds retain their wings they are are also buoyant. Water-dispersed seeds have probably given rise to populations of *B. alpina* found along major rivers and lowland lake shores which drain from the plant's alpine and sub-alpine habitats.

Cerastium nigrescens (H.C. Watson) Edmondston ex H.C. Watson

SHETLAND MOUSE-EAR

Cerastium nigrescens (Shetland mouse-ear) was brought to the attention of botanists by the brilliant young Shetlander Thomas Edmondston. In 1837, when only eleven years of age, he discovered the plant on the Keen of Hamar, a hill of serpentinite rock situated on the north-east coast of Unst, the most northerly island of Shetland.

A vigorous exchange of views relating to the taxonomic distinction of *Cerastium nigrescens* later took place between Edmondston and the eminent Hewett Cottrell Watson. Edmondston first thought that the plant was the true *Cerastium latifolium* named by Linnaeus, and that similar plants from the mountains of mainland Scotland and Wales, that British botanists regarded as *Cerastium latifolium* of Linnaeus, were in fact varieties of the alpine mouse-ear *(Cerastium alpinum)*. Watson strongly disagreed with Edmondston and doubted if the Shetland mouse-ear was any more than a mere form or variety of Linnaeus' *Cerastium latifolium* claiming that Edmondston misunderstood this group of plants. The young botanist vigorously upheld his views but then appeared to change his mind: he began to think that his Shetland mouse-ear was actually a separate species from Linnaeus' *Cerastium latifolium* and started to use the name *Cerastium nigrescens* although failed to follow the strict rules that govern the publication of new plant names. Although Edmondston was later to defer to the authority of Watson (as he explained in his *Flora of Shetland* in 1845), Watson had obviously been impressed with the spirited young 17-year old and after a chance encounter in Braemar in 1844 was to befriend and support him. Since the time of Thomas Edmondston there have been continual differences of opinion about the taxonomic status of the Shetland mouse-ear and it has been recognized as a form, variety, subspecies and separate species by various botanists. Although disagreement still remains, few botanists have any reservations about the interest and conservation importance of the plant. It is clearly closely related to Arctic mouse-ear (now *Cerastium arcticum* Lange but referred to as *C. latifolium* in Edmondston's time) but differs from it mainly in its rounded leaves, short broad capsules, broad sepals at fruiting stage and dense short glandular hairs. In its wild habitat *Cerastium nigrescens* is dark purple-tinged (hence nigrescens), which is not only distinctive but attractively contrasts with the large white flowers produced mainly in June. The plant is endemic to the island of Unst.

Cerastium nigrescens, Shetland. (Photo: P Lusby).

Although the Keen of Hamar reaches an altitude of only 89m, the extensive areas of sparsely colonized stony soil (often referred to as serpentinite debris) and relentless wind, impart an atmosphere of a high mountain plateau. *C. nigrescens* is restricted to the most open areas of debris and is not found in the wind-clipped sedge-grass-heath and heath vegetation that forms most of the remaining vegetation of the Keen of Hamar which has not been modified by agriculture. Other rare and scarce plants that accompany *Cerastium* include Norwegian sandwort (*Arenaria norvegica* subsp. *norvegica*), and northern rock-cress (*Cardaminopsis petraea*) with common species such as thrift (*Armeria maritima*), sea plantain (*Plantago maritima*), kidney vetch (*Anthyllis vulneraria*), sea campion (*Silene uniflora*), moss campion (*Silene acaulis*), red fescue (*Festuca rubra*), brown bent (*Agrostis vinealis*) and wild thyme (*Thymus polytrichus* subsp. *britannicus*).

The open serpentinite gravel has several features that may account for the very sparse colonization by plants. Like many soils derived from serpentinite rock it contains high levels of metals such as magnesium, chromium and nickel. However, the physical structure of the rock has been emphasized as the most likely cause of the lack of vegetation on the Keen of Hamar debris. Weathering causes the rock to fracture into small fragments which form a very well-drained soil. In periods of dry weather in summer, drought conditions can prevail, subjecting the plants to quite severe water stress. In winter, frost heave, coupled with strong winds, can uproot plants. *Cerastium nigrescens* displays several adaptations to dry and disturbed conditions, including dense glandular hairs, rather fleshy leaves and a strong and extensive root system. Even so, it is common to see plants that have been severely dislodged during the winter with most of the root system above ground, so that they are more appropriately described as tethered rather than rooted in the soil!

On the adjacent hill of Muckle Heog a much smaller population of *Cerastium* occurs in wetter, flushed serpentinite debris, and here the plants do not display some of the distinctive features as those on the Keen of Hamar to such a degree; the leaves tend to be less fleshy and less rounded with more acute tips, and the petals are slightly narrower. Thomas Edmondston mentioned a variety *(acutifolium)* that grew more inland and there is little doubt that it was the Muckle Heog plants to which he was referring. There has been some speculation that these plants are the result of a cross between the common mouse-ear *(Cerastium fontanum)* and *Cerastium nigrescens*, but response to wetter growing conditions is a more likely explanation.

In 1967 a considerable area of the Keen of Hamar was fertilized and reseeded. This converted the natural vegetation to pasture. Fortunately most of the remaining area that contains *Cerastium nigrescens* is now within the boundaries of a National Nature Reserve and a Site of Special Scientific Interest.

Many of the people of Unst value the Keen of Hamar and its interesting flora and although they may not be familiar with the botanical details and ecology of the hill, most know of the Shetland mouse-ear which they fondly and aptly refer to as 'Edmondstonian', after their long-lost botanical prodigy who put Shetland firmly on the botanical map.

Cicerbita alpina (L.) Wallr.
ALPINE SOW-THISTLE

Alpine sow-thistle is one of several discoveries made by George Don on Lochnagar. The north-east corrie of this mountain supports Britain's largest population of *Cicerbita alpina*, and it was probably on the broad, sloping ledge at about 1000m that Don found the plant in 1801.

The Aberdeenshire naturalist, William MacGillivray, visited Lochnagar in 1850 with two "sedate elderly gentlemen, little apt to be sent astray by pulses of enthusiasm", who nevertheless clung "with their hands to the face of a precipice 500 feet high ... [and] having found a shelf that led into the the great fissure ... brought [out] a bunch of the *Mulgedium*" [*Cicerbita alpina*]. MacGillivray referred his companions' behaviour to 'Phytomania', a condition that was regretably common among botanists of the 19th century, especially where rare and visually impressive plants were concerned. Although collecting was frequent at some of *Cicerbita*'s sites, notably Glen Doll, written accounts suggest there were as many failed collecting attempts due to the inaccessibility of the plant's sites. Most collectors seemed content with (or were only able to grab with one free hand) a flowering stem and leaf, rather than uprooting the specimens. At one site where the species has knowingly disappeared, Glen Canness, its final demise was probably the hot, dry summer of 1976. Nevertheless, alpine sow-thistle is one of our rarest mountain plants, present in just five sites in the Clova-Caenlochan area of the east-central Highlands.

Cicerbita alpina, Angus,
17 viii 1995.

Lochnagar, better known to most for its Royal patronage than for its botanical richness, has been described as the most rewarding among the "floristically dull acid mountains" of the north. Although the parent rocks of all the Scottish sites are acidic, base-rich intrusions are contained within them and *Cicerbita* seems to grow in association with these. On Lochnagar the plants grow on or near to fault zones where the typical pink or grey granite is replaced by cream or greenish heavily weathered rock in which the basic minerals epidote and calcite are present. Typical calcicoles among *Cicerbita*'s associates indicate local base-enrichment: alpine mouse-ear (*Cerastium alpinum*), frog orchid (*Coeloglossum viride*), yellow and purple saxifrages (*Saxifraga azoides* and *S. oppositifolia*) and holly fern (*Polystichum lonchitis*).

Cicerbita alpina is predominantly found on north-facing mountain ledges or in ravines which are humid and sheltered from strong winds. Snow lies late in these localities and melt-water provides a valuable source of moisture when the shoots emerge in late spring to early summer. Ledges and gulleys are also receiving sites for mineral-debris and nutrient-enriched seepage water from slopes and rocks above. This constant percolation and occasional rock-falls combine to counteract acidification of the soil and leaching of nutrients.

In its surviving sites *Cicerbita alpina* is for the most part protected from grazing, an essential requirement given the extreme palatibility of the tall leafy shoots. However, at Caenlochan deer occasionally succeed in reaching the single ledge on which the plant grows, aided in some years by a bank of snow. A well-trodden deer track passes just below the ledge and is perhaps testimony to many frustrated attempts. The present size of the deer and sheep population in the eastern Highlands makes the future of *Cicerbita* in Scotland less than certain. Exclosures are the best short-term measure for protecting individual populations but, in the longer term, allowing recovery of the plant's natural habitat is the best way to ensure its survival.

Outside Scotland, *Cicerbita alpina* occurs in all the main mountain areas of Europe; from the Pyrenees to the Alps, the Apennines and the Balkans, and throughout the Scandinavian mountains. It is typically a plant of montane tall-herb vegetation, described by Linnaeus in his *Flora Lapponica* of 1737 and by many botanists over the last two centuries. The eminent Norwegian botanist, Rolf Nordhagen, recognized *Cicerbita alpina* as so characteristic of this important plant community (an alliance), he named it *Lactucion alpinae* (now *Cicerbition alpini*).

This tall-herb community is represented as alpine meadow, low-alpine willow scrub or as an understorey of sub-alpine forest, most commonly birch, or sparse Norway spruce. As meadow or woodland field-layer, the community is typically very luxuriant, reaching two metres in height. Amongst the associates of *Cicerbita alpina* common to Norway and Scotland are: melancholy thistle (*Cirsium heterophyllum*), marsh hawk's-beard (*Crepis paludosa*), meadowsweet (*Filipendula ulmaria*), water avens (*Geum rivale*), wood crane's-bill, (*Geranium sylvaticum*), alpine saw-wort, (*Saussurea alpina*) and globeflower (*Trollius europaeus*). Dominant or co-dominant with *Cicerbita* in Norway, but absent from Scotland, is the tall northern monkshood (*Aconitum lycoctonum*). Also frequent in the *Cicerbition alpini* in Scandinavia, but very rare in Scotland, is *Polygonatum verticillatum*.

The development of this tall-herb vegetation is primarily dependent on high precipitation and humidity, and a relatively fertile mineral soil. This fertility may derive from underlying base-rich rocks, or in areas of acidic geology, from the addition of nutrients in snow-melt, run-off and moving ground-water, particularly if it flows over basic intrusions. Such waters irrigate scree slopes, cliff-bases, gully mouths, and other moist, sloping sites.

Cicerbita alpina, Angus, 17 viii 1995.

The more or less isolated occurrences of *Cicerbita alpina* in the alpine zone in Scandinavia (as opposed to in sub-alpine forest), can be compared with the fragmentary distribution of the species in Scotland. The Scandinavian populations above the tree-line are believed to be relics from a time when forests reached higher altitudes during the post-glacial warm Atlantic period. Any seeds of *C. alpina* which now reach the alpine zone rarely establish due to the generally harsher climate. For the same reason, reproduction of established alpine populations is very infrequent. The Scottish populations on montane cliffs are like these alpine relics in their small size and isolation. Both are subject to greater extremes of climate and a higher level of disturbance than are populations in sub-alpine forest. Above the tree-line *C. alpina* is confined to sites providing sufficient moisture and shelter such as moist meadows, fast-flowing stream sides, slanting depressions and rock bluffs.

Cicerbita alpina has formerly been placed in three different genera: *Sonchus, Mulgedium* and *Lactuca. Mulgedium*, from the Latin *mulgeo*, to milk, and *Lactuca*, from *lac*, milk, refer to the milky-white sap contained in the plant. *Cicerbita* is derived from cicharba, a name used for any plant resembling the sow-thistle (*Sonchus*). There is a popular belief that the plant, if eaten, increases milk-production in cattle and nursing-mothers. Linnaeus recorded the Laps ate the young, skinned shoots as a delicacy, though he found this rather bitter. The plant contains sesquiterpenes and furanocoumarins which are insect-deterrent chemicals. Although bitter-tasting, they appear to have little effect on mammals. In Scandinavia it is known as bear blue-lettuce, pig-grass and alpine salad. One patch of *C. alpina* on Lochnagar became known to rock climbers as the "potato-patch" not, one assumes, for any edible likeness.

Its striking blue flowers, opening in late July to August, have given it the name alpine chicory. They are short-lived and close in the evening like those of the dandelion and many other composites. Butterflies and bumble-bees are the chief pollinators. Although seedlings have been found in some of the Scottish populations, seed-set appears to be somewhat irregular and the seed produced is not always viable. Limited opportunities for cross-pollination and the vagaries of the montane climate may be partly responsible for this.

Cornus suecica L.

DWARF CORNEL

Dwarf cornel was discovered on The Cheviot in Northumberland by the Rev. Dr Thomas Penny in the latter half of the 16th century. The plant was recorded in *Rariorum Plantarum Historia* of 1601 by Clusius who acknowledged a communication from Penny. Both botanist and entomologist, Penny was said to be "so learned" in plants that he was called the "second Dioscorides", and is known to have corresponded with some of the most well-known botanists of his day, among them Conrad Gesner, Matthius de L'Obel and John Gerard. *Cornus suecica* was not recorded north of the border until John Lightfoot, accompanied by Thomas Pennant, found it at the head of Little Loch Broom in West Ross during his tour of Scotland in 1772.

Cornus suecica, Perthshire, 28 vi 1995.

The Cheviot is one of the few English localities for dwarf cornel, but the plant also occurs rarely in Lancashire and north-east Yorkshire. In southern Scotland it is scarce in the Moffat Hills and Peeblesshire, and only becomes common in the central and western Highlands. It is absent from the Scottish islands with the exception of Orkney and Shetland.

Cornus suecica has an arctic-subarctic distribution with a southern European limit in northern Germany. From the Netherlands and Denmark it spreads north through Scandinavia into western Russia. It fails to reach the continental interior of Asia, but reappears in northern Japan. Westwards, its range encompasses the coastal districts of North America and Greenland, Iceland and Faeröe.

In Britain dwarf cornel is a plant of montane wet heaths, usually on acid rocks where it frequently marks the altitudinal transition between upland heaths and mires, and montane vegetation of sub-shrubs, lichen or moss heath. Blaeberry (*Vaccinium myrtillus*) frequently dominates this vegetation zone, often concealing *C. suecica* under its canopy. Throughout its range, cloudberry (*Rubus chamaemorus*) is a characteristic associate, sharing a very similar sub-oceanic distribution. *C. suecica* is intolerant of extreme dryness and winter cold which are generally avoided in the oceanic north and west of its distribution, but which restrict the plant to shady north- and east-facing slopes and hollows in the less oceanic south and east. Although a calcifuge, the plant prefers a sub-soil rich in humus and permanently moist. Where rainfall does not provide adequate moisture, melt-water from late-lying snow on lee slopes and in hollows becomes a vital seasonal source.

Although the heath community in which *C. suecica* occurs over much of its range in Britain may be the climax vegetation, near its lower altitudinal limits it falls within the range of historical pine-birch forest. In Scandinavia the plant grows naturally at or just above the margin of sub-alpine birch wood in an assemblage containing several species common to the Scottish sites: chickweed wintergreen (*Trientalis europaea*), common cow-wheat (*Melampyrum pratense*), goldenrod (*Solidago vigaurea*) and crowberry (*Empetrum nigrum* subsp. *hermaphroditum*). Above the tree-line, in the low-alpine zone it is asssociated with blue heath (*Phyllodoce caerulea*), another plant of late-lying snow hollows. On the Sow of Atholl in Perthshire these two species grow in close proximity.

Finding dwarf cornel may require some careful searching as the species does not flower very freely in Britain and is often rendered inconspicuous by its habit of sheltering beneath bilberry or heather. However, one is commonly rewarded by quite large patches which are formed by the plant's slender underground rhizomes. In flower, *C. suecica* is an ideal subject for black and white photography. The four white or pale yellow 'petals' are actually bracts, in the centre of which are eight to 25 tiny flowers with dark red to purple-black petals. Little is known about what insects are attracted by this monochrome inflorescence, though hover-flies are among the observed potential pollinators.

As the season advances dwarf cornel produces a display of glorious multicolour. The leaves turn yellow and shades of red and purple, providing a marbled backdrop for the bright red fruits. It is these which give the plant its alternative common name, dwarf honeysuckle.

John Lightfoot recorded in 1777 that the fruits had "a sweet, waterish taste and are supposed by the Highlanders to create a great appetite, whence the Erse [Gaelic] name of the plant", lus-a-chraois, plant of gluttony. In Norway dwarf cornel is known as skrubbaer, wolfberry and in Sweden, hönsbär and chickenberry. Birds such as grouse are known to eat the fruits and thus disperse the seeds. The yield of fruit is usually rather poor and the flesh somewhat insipid, but when stewed with bilberries and other wild fruits, dwarf cornel makes a pleasant appetizer or dessert. Eskimos and Laps harvest the fruits for winter food and make them into tarts.

Cystopteris dickieana Sim

Dickie's Bladder-fern

Dickie's bladder-fern is arguably the rarest fern in the British Isles but there is intense debate among fern specialists (pteridologists) as to the correct taxonomic rank of the plant. The problems associated with this fern raise interesting questions relating to the conservation of other rare plants.

The discoverer of the plant in a Kincardineshire sea cave was William Knight (1786–1844), Professor of Natural Philosophy at Marischal College, Aberdeen. Although he drew the attention of his pupils to this unusual fern, it was George Dickie (possibly one of these pupils) who first published a record of the plant in his *Flora Aberdonensis* of 1838. At this stage Dickie made no move to recognize the fern as a separate species but referred to it as *Cistopteris* (sic) *fragilis*. Among the pteridologists that first received living specimens of the fern from Dickie was the nurseryman Robert Sim of Foot's Cray in Kent. In a short paper in the *Gardener's and Farmer's Journal* of 1848 Sim described the fern, including the spores, and considered it a new species : *Cystopteris dickieana*, in honour of the man who brought the plant to the attention of the botanical world. Sim's opinion of the plant was that "If any of the recorded species of *Cystopteris* apart from *C. fragilis* have claim to rank as such, so also must *C. dickieana*". In 1849 Thomas Moore, a well respected pteridologist, had some sympathy with Sim's suggestion that Dickie's bladder-fern might be considered a species but was inclined overall to "... consider it as a marked variety of *C. fragilis* ...".

Other botanists of the time were intrigued by the beauty and distinctness of the fern but still tended to consider it a variety of *Cystopteris fragilis*. It is really only since the 1930s that the standard British floras have consistently recognized *Cystopteris dickieana* as a separate species. Dickie's bladder-fern became widely known to fern fanciers by the early 1850s when the Victorian fern craze was approaching its height. Such a distinct and easily grown fern, which retains its distinguishing characteristics in cultivation, attracted particularly heavy collecting until Dickie reported it completely extirpated from its original site in 1860. However, considerable mystery surrounds his statement as there are herbarium specimens in existence dated 1871, 1879 and 1893 whose labels suggest they were collected from the original locality. In fact today the cave contains a population in excess of one hundred plants. Whether the fern managed to recolonize its original cave from others to the north where it was known to have persisted or whether Dickie did not actually make the observation himself, is not known. Recent research has demonstrated that many ferns, including *C. dickieana*, form spore banks and it is possible that the population regenerated in this way.

The broad-mouthed, east-facing caves (aptly known as 'yawns') in which *C. dickieana* grows along the Kincardine coast, are very atmospheric. Except in the driest conditions they are constantly humid and the fern grows mainly in a single roof fissure accompanied by a few mosses and three other ferns, lady-fern (*Athyrium filix-*

femina), sea spleenwort (*Asplenium marinum*) and broad buckler-fern (*Dryopteris dilatata*). Like the very closely related *C. fragilis*, *C. dickieana* is confined to base-rich rocks and the interior of its cave roof is composed of schist containing lime-rich veins which also receive seepage water from basic lavas above.

The features that distinguish typical *Cystopteris dickieana* from *C. fragilis* are the broader, less-divided and more closely spaced pinnae which gives a crowded appearance to each frond, and also the ultimate divisions of the latter are blunter. Microscopically, the surface of the spores of *C. dickieana* appears granular and the outer spore wall is wrinkled and ridged (rugose) in contrast to the spiny or 'echinate' spores of *C. fragilis*. It is this spore character that has assumed most importance in the identification of *Cystopteris dickieana*. Recently, plants from Perthshire with fronds similar to *C. fragilis* but with wrinkled spore surfaces have been identified as *C. dickieana*. Further, plants of *C. dickieana* with the granular spore surface but with smooth, non-wrinkled spore walls have been recorded. Population studies in America and Scandinavia have also shown that the diagnostic spore and frond characters of *C. dickieana* do not always go hand-in-hand. In various populations plants have been found with non-spiny and spiny spores which are more similar to each other in frond structure than to plants with their respective spore types in other populations. Therefore variation in frond shape does not always support the assumption that *C. fragilis* and *C. dickieana* can be separated by spore characters also. Recent work involving the identification of certain proteins in populations of *Cystopteris* (including the Kincardineshire plants) has revealed considerable variation between populations. In short, there does not appear to be any overall correlation between frond shape, spore type and protein (allozyme) structure among populations of what are called *C. dickieana* and *C. fragilis*. From this wider and more detailed analysis there is a growing opinion among pteridologists that the original Kincardineshire cave plants may be best considered as a distinct populational variant of *C. fragilis*, rather than a separate species. If this were to happen the variation to be included within *C. fragilis* would be widened. Whatever the outcome, the combination of distinctive characters of the Kincardineshire plants is very rare in terms of biological diversity and should receive appropriate conservation regardless of taxonomic rank. The constancy of the characteristic features of the Kincardineshire plants in cultivation (even when grown in open, well lit conditions), together with the relative ease of cultivation, may have been significant for the survival of the plant at the height of the fern craze when the demand for it could have been met from material propagated from spores.

Plants identified as *C. dickieana* have been recorded from a wide area and the total distribution is considered circumpolar with a southerly extension to northern Africa and the northern Andes in South America.

Cystopteris dickieana growing in a dark sea cave, Kincardineshire, 3 viii 1994.

Throughout its range *C. dickieana* is essentially a montane species confined to the uppermost forest belts. C. N. Page, in *The Ferns of Britain and Ireland*, points out that although the coastal habitat of the Kincardineshire plants may at first seem unusual, other montane plants such as *Oxytropis halleri* are known to behave similarly and it is thought that it is the relatively cool summers on the coast that mirrors the montane climate. The coastal plants must also be able to withstand salt-laden air which is of particular interest as it has been suggested that this adaptation may have allowed *C. dickieana* to survive in a relatively pure form in the Kincardineshire caves after possible displacement by *C. fragilis* or hybrids in more montane sites in Scotland. This begs the question whether plants with the spore type of what is called *C. dickieana* but with the frond structure of *C. fragilis* are hybrids between two separable species. In southern central Norway, populations of plants identified as mixtures of *C. dickieana* and *C. fragilis* contain plants with many abortive pollen grains which may suggest hybridization. Further, other experimental crosses made between the Kincardineshire *C. dickieana* plants and *C. fragilis* from eastern Europe have resulted in sterile hybrids. This apparent reproductive isolation suggests that we should not be too ready to consider them as one species, although a larger programme of crosses between *C. dickieana* and *C. fragilis* needs to be carried out before a concensus of opinion can be reached.

Diapensia lapponica L.

DIAPENSIA

Diapensia lapponica is the epitome of an arctic-alpine plant, its cushions bearing abundant showy white flowers under the most extreme conditions of cold and exposure. In Scotland, the rarity, attractiveness and normally free-flowering nature of the plant has made it a popular focus for annual pilgrimages among mountain botanists.

Diapensia was one of the major additions to the Scottish flora in the 1950s being discovered near Glenfinnan in Westerness in July 1951 by Mr C. F. Tebbutt during an ornithological excursion. This site remains the only known locality but the excitement of finding *Diapensia* has led to a number of claims of a second site, all so far having turned out to be misidentifications. The scarce white form of the pink-flowered trailing azalea (*Loiseleuria procumbens*) has been mistaken more than once for *Diapensia*. A clear diagnostic character of the flowers of *Diapensia* is the broad, flat anther filaments which are joined to near the top of the petals. When the plant is not flowering it can easily be distinguished from *Loiseleuria procumbens* and crowberry (*Empetrum nigrum* subsp. *hermaphroditum*) by the tight rosettes of paddle-shaped (obovate) leaves. Those of *Empetrum* and *Loiseleuria* are never obovate (wider above the middle) but are linear-oblong to oval-oblong.

Diapensia grows on an exposed summit ridge but the precise area over which it occurs contrasts with the rest of the area by the upturned and convoluted rock outcrops. The main associated species are woolly hair-moss (*Racomitrium lanuginosum*), crowberry (*Empetrum nigrum* subsp. *hermaphroditum*), pill sedge (*Carex pilulifera*) and mat-grass (*Nardus stricta*) with others such as *Loiseleuria procumbens*, viviparous sheep's-fescue (*Festuca vivipara*) and alpine lady's-mantle (*Alchemilla alpina*).

Diapensia lapponica is divided into two subspecies, subsp. *lapponica* and subsp. *obovata*, based, as the name suggests, primarily on the more markedly obovate leaves of the latter. There is a general separation in the distribution of the two subspecies. Subsp. *lapponica*, which includes the Scottish population, occurs on both sides of the Atlantic Ocean ranging in eastern North America, from Mt Washington in New Hampshire to Ellesmere Island, to Greenland, Iceland, Britain, Scandinavia and the western former Soviet arctic and subarctic. Subsp. *obovata* is found in eastern arctic and subarctic Siberia, Korea, Japan, Aleutian Islands, Alaska and the Yukon. Taken as a whole the species has a circumpolar distribution. Pink-flowered forms of subsp. *obovata* have been named forma *rosea*.

Diapensia lapponica is the archetypal stress tolerator and avoids competition by growing in the most inhospitable places. Its typical habitat is very cold and exposed areas on mountains or open tundra which are blown free of snow in winter and where the soils are subjected to frost disturbance. The plant even survives on Mt Washington where the world's highest windspeed of 234 miles per hour was recorded. However, the greatest test for any plant growing in such exposed places are the extremes of temperature and moisture levels it must tolerate during the year. The exception to the montane and tundra habitats of *Diapensia* is the foggy coastal locations

Diapensia lapponica, Inverness-shire, 14 vi 1995.

of southern Newfoundland where a 'pseudoarctic' climate is produced by mixing of the ice-laden Labrador current and the warmer Gulf Stream. *Diapensia* is generally a plant of acid, nutrient-poor soils and avoids areas of base-rich rocks but is known to grow in one area on serpentinite in Newfoundland which, being high in magnesium and often other metals, is toxic to many plants.

To survive the rigours of the climate it is to be expected that *Diapensia* exhibits several adaptations. The cushions trap and derive nutrients from wind-blown particles and also act like greenhouses in that they absorb solar radiation and minimise heat loss. It has been found that on sunny spring days from morning until mid-afternoon the leaf temperature of *Diapensia* may be higher than anything else in the vicinity. This ability to trap heat extends the otherwise very short growing season. The leaf cuticles, which are extremely resistant to the passage of water, enable *Diapensia* to survive in highly

Diapensia lapponica,
Inverness-shire, 31 v 1995.

drought-prone areas in winter. However, in summer, restricted water passage can be a disadvantage as leaf cooling by transpiration cannot take place effectively. To counter this *Diapensia* has a fairly high heat resistance in comparison with many arctic-alpine plants. The frost hardiness and drought resistance of *Diapensia* have been shown to change seasonally to coincide with periods of greatest stress. Drought resistance is high from late autumn to the beginning of May but during this month tolerance decreases sharply and the plant is most drought sensitive in early July. In the Arctic thawing occurs later than at lower latitudes. In Scotland, near the edge of its geographic range *Diapensia* may well be extremely vulnerable to drought stress in warm, dry periods in late May, June and July. In the warm June of 1983 it was noted that many of the flowers of the Scottish population had withered prematurely, some in bud. The distribution of *Diapensia* is limited to areas with cool summers.

The flowering period is usually late-May to mid-June but can vary considerably with the season. The flowers produce nectar at the base of the shallow corolla tube which is readily accessible to the few potential pollinators. The closely-packed flowers enable the main pollinators, bumble-bees, to crawl over them without having to frequently take-off and re-land into the wind. Ants have been recorded on *Diapensia* flowers but, because of the lack of body hairs, they are not efficient pollinators. There has been doubt as to whether Scottish plants produce viable seed but at least one specialist alpine grower has succeeded in raising plants from wild-collected seed. However, a licence is required from Scottish Natural Heritage for the collection of seed from this or any other plant listed in Schedule 8 of the Wildlife and Countryside Act 1981.

Dryas octopetala L.

MOUNTAIN AVENS

The first record of *Dryas octopetala* growing wild in the British Isles is from western Ireland, "in the Mountaines betwixt Gort and Galloway [Galway]" where it was found by the Rev. Richard Heaton, probably on his way to his Clare parish, in the 1630s. *Dryas* is a very old member of our flora; remains have been recovered in deposits in southern England dated to the Full-glacial (about 20,000 years ago). The Late-glacial

(10,000–15,000 years ago) is known as the Dryas period because of the abundance of fossil *Dryas* records. Many of these are from south of the species' present distribution in Britain and Europe. In southern England and in south and east Ireland, *Dryas* grew in a tundra-like vegetation with other arctic-alpines, including *Alchemilla alpina*, *Minuartia sedoides*, *Draba incana*, *Salix reticulata*, and *Potentilla fruticosa*. Elsewhere in the lowlands, for example on the Isle of Man, Post-glacial finds indicate the persistence of a treeless vegetation for varying lengths of time after the ice-retreat, depending on climate and the type of substrate.

It is fitting that *Dryas octopetala* should have been first recorded from the Burren; the abundance of the plant is one of the celebrated features of the limestone of counties Clare and Galway. It has been locally christened the Burren Rose, and one botanical traveller in 1808 thought "it might almost be called Burren grass, it constitutes so large a proportion of the whole vegetation". Outside the Burren there are scattered localities near the coast in Donegal, Londonderry and Fermanagh, in the northern Pennines and the Lake District, and in Snowdonia. In Scotland, *Dryas* is most frequent on the north coast and on calcareous rocks in mountains from Angus to Argyll. Its total range extends from Kintyre to the Inner Hebrides and the north coast of Sutherland.

The distribution of *Dryas* is tied to the occurrence of basic rocks which include Carboniferous limestone (north Yorkshire and Ireland), metamorphosed (sugar) limestone (Cairngorms and Teesdale), mica-schists (central Grampians), basalts (Skye) and lavas (Lake District and Wales). On the north coast it occurs on the Durness limestone and calcareous shell sand. As well as being a typical calcicole in Britain, *Dryas* is limited to areas with a sufficiently cool and humid climate. South of the Border and in Ireland, all localities fall within the 25°C isotherm of mean annual

Dryas octopetala, Sutherland, *16 vi 1996.*

maximum temperature, and in Scotland, within the 23°C isotherm. Rainfall for its British distribution exceeds 1000mm per year. A considerable proportion of the total precipitation falls as snow in its montane localities, although in Britain, *Dryas* is restricted to exposed sites relatively free of snow cover. In its lowland localities it grows on open slopes with free-drainage or on cliffs. Such sites, and the sand dunes of its coastal habitat, are kept open and base-rich by erosion and falling rock debris or accretion of sand.

Linnaeus' type locality for *Dryas octopetala* reads, "*Habitat in Alpibus Lappoicus, Heleveticus, Austriacis,*

Sabaudicis, Hibernicis, Sibiricis", although he undoubtedly knew the plant best from the Swedish mountains. The vernacular name 'polymountain' is very appropriate for *Dryas* as it occurs in all the mountain ranges of Europe. North of the Arctic Circle it is circumpolar, though in North America it extends down the Rocky Mountains to Colorado, and in Asia it reaches south to Mongolia, North Korea and Japan.

Not surprisingly for such a widely distributed plant, *Dryas* shows considerable variation, particularly in leaf form. Attempts have been made to distinguish various forms as subspecies or even separate species, but it seems at present to be best regarded as one polymorphic species. British plants with black instead of red sepal hairs, and lacking scales on the leaf stalks and midribs have been described as *Dryas babingtonia* representing a western European race endemic to Scotland, Ireland and Norway. However, this form can occur together with 'typical' *D. octopetala* in British populations. Ecological specialization is perhaps most evident (or most studied) in Alaska where subspecies *octopetala* and a second subspecies, *alaskensis*, occur in fellfield and snowbed communities respectively. Differences in leaf size and growth habit are clear adaptations to their respective habitats; snowbed plants have large, evergreen leaves and grow in a prostrate, spreading manner, whereas fellfield plants have small deciduous leaves and form dense mats. Fellfield plants are also less tolerant of shading which they would not normally experience in their more exposed habitat.

The large amount of variation shown by *Dryas* is of interest to the taxonomist and is much valued by gardeners. Reginald Farrer, plant-hunter and rock gardener, regarded *Dryas octopetala*, "the sovereign of alpine shrubs". Even without the wide choice of cultivars that are available, *Dryas* has much to recommend it for the rockery or alpine garden. Its leaves, deep green above, silky-white below, are scalloped like tiny oak leaves, whence the name *Dryas* [Greek for oak]. The constant glow of the flowers on a sunny day is not an illusion, but an adaptation known as sun-tracking whereby the dish-shaped flowers constantly turn to face the sun throughout the day. Insects are attracted into the heated interior of the blossoms. Although usually eight-petalled, as its specific name suggests, *Dryas* flowers are frequently found with more, both in the garden and in the wild, although true doubles are rare. Beauty does not pass with the withering of the flowers; they are succeeded by seed heads of silky plumes like those of pasque-flower, *Pulsatilla*, but are characteristically twisted in *Dryas*. To add to these virtues, the plant is completely frost resistant, easy to propagate by layering or cuttings, retains its leaves in winter and keeps weeds at bay by its mat-forming habit.

Dryas vegetation in Britain and Ireland is quite varied, mainly reflecting the range in altitude (sea level to 1050m on Ben Avon) and local soil drainage. Two *Dryas* communities are recognized in Britain; one a montane community found on crags and ledges, the other a sub-montane grass-heath, descending to the coast in the far north. Many stands of both communities are small in extent and more or less isolated. Grazing by sheep and rabbits is partly to blame for this; vegetation containing *Dryas* is always most luxuriant and the plant more floriferous when ungrazed. However, it is also true that a long history of grazing and scrub-clearance in some areas has kept the vegetation open and the soil thin, favouring *Dryas*; this is certainly the case in the Burren.

Eriocaulon aquaticum (Hill) Druce

PIPEWORT

Eriocaulon aquaticum is the only European representative of the largely tropical and sub-tropical family Eriocaulaceae. The interpretation of its peculiarly widespread occurence in north-eastern North America, whilst restricted in Europe to the western seaboard of Scotland and Ireland has generated considerable botanical interest and controversy.

James Robertson's collection from Skye in 1768 is generally accepted as the first record of the plant in the British Isles. However, a footnote in W. J. Hooker's *Flora Scotica* of 1821 records a communication from a Mr Maughan who had found a note attached to a specimen of *E. aquaticum* in the herbarium of the late Dr John Walker of Edinburgh. This note claimed that the specimen had been collected by Sir John Macpherson on 11 September 1764 from a small loch by the road from Sconsar to Giesto on Skye. When he noticed it he apparently "... leaped from his horse, waded into the lake and brought it out ...".

Eriocaulon aquaticum, Skye, 4 viii 1995.

John Hope made a detailed description of the Skye *Eriocaulon*. Because Linnaeus' general description of the genus did not include some of the characters of the Skye plant, Hope thought that it may be best to consider it a new genus. He proposed to call it *Nasmythia articulata* in honour of Sir James Nasmyth (or Naesmyth), who studied under Linnaeus in Sweden. Unfortunately Hope sent the description, as well as a beautiful illustration of the plant by James Robertson, to the Earl of Bute in whose library it was seen by Sir John Hill. In an attempt to make his forthcoming *Herbarium Britannicum* of 1769 as comprehensive as possible, Hill included Hope's information but without his consent and, further, named the plant *Cespa aquatica*. Linnaeus later amended his description of the genus *Eriocaulon* to include the characters of the Skye plant and both he and Hope identified it as *Eriocaulon decangulare*. However, William Withering considered it a separate species from *E. decangulare* and named it *E. septangulare* in his *Botanical Arrangement of all the Vegetables Naturally Growing in Great Britain* of 1776. The celebrated botanist G. C. Druce later pointed out that the earliest published specific name must be used which was Hill's name of *aquatica* unfortunately making another name change necessary.

In Scotland, *Eriocaulon aquaticum* has its greatest concentration on Skye but with some large populations on Coll and the Ardnamurchan Peninsula in Argyll.

There is a single recent record from Tiree but the plant has not been re-recorded from Iona where it was once known. In Ireland its headquarters are in Connemara but it has been recorded in scattered localities from western Cork to west Donegal. Throughout its range *E. aquaticum* occurs in the more sheltered margins of peaty lochans but the plant seems to tolerate a little base enrichment. The length of the flowering stem (or scape) varies with the depth of water and in Scotland stems two feet long have been recorded, whilst in Ireland they may reach three feet. Population size is very variable, ranging from just a few plants to many thousands. Commonly associated species are many-stalked spike-rush (*Eleocharis multicaulis*), water lobelia (*Lobelia dortmanna*), lesser spearwort (*Ranunculus flammula*), bottle sedge (*Carex rostrata*), bogbean (*Menyanthes trifoliata*), floating club-rush (*Eleogiton fluitans*), bulbous rush (*Juncus bulbosus*), broad-leaved pondweed (*Potamogeton natans*) and white water-lily (*Nymphaea alba*). On Coll, the local intermediate sundew (*Drosera intermedia*) often occurs in the vicinity of *Eriocaulon aquaticum*, and on Skye and Coll it is sometimes accompanied by the scarce bog-hair grass (*Deschampsia setacea*).

With a few other species in the British flora, such as Irish lady's-tresses (*Spiranthes romanzoffiana*), blue-eyed grass (*Sisyrinchium bermudiana*) and slender naiad (*Najas flexilis*) it forms the North American Element. This group is characterized by their very unequal distribution on either side of the Atlantic with their greatest concentration in America and a very restricted occurrence in Europe. The British Isles represents the sole native European locality for *Eriocaulon aquaticum* although it has become naturalized in the Azores. In North America its main area extends along the east coast from Nova Scotia to Pennsylvania but reaches north and south to Newfoundland and Virginia respectively. To the west it reaches at least as far as North Dakota. Several theories have been put forward to explain the current distribution of *E. aquaticum*.

The pollen of *Eriocaulon* is quite distinctive and has been found in Irish deposits dating from the early Atlantic period (c.7000 years ago) and probably from the Hoxnian or 'Great Inter-glacial' period some 700,000 years ago. All pollen from these deposits matches that of modern Irish plants whilst those from America have smaller pollen grains. It has also been shown that Irish plants have twice the number of chromosomes compared with American plants. These observations suggest that the populations of *Eriocaulon* on either side of the Atlantic have been separated for a very long time and does not agree with a theory that the plant has more recently reached the British Isles from America by way of bird migration. There is generally more support for considering the present distribution of *E. aquaticum* as representing a relictual fragment of a once more continuous distribution. Some biologists think that *Eriocaulon aquaticum* migrated across a land connection between Britain and America in Tertiary times (c.1.6–65 million years ago), whilst others favour a more continuous Pre-glacial distribution in the other direction across Siberia and Alaska. If, like other species with more continuous present day northern or boreal distributions, *Eriocaulon* survived the last glaciation south of the ice sheet it is unclear why it has been so unsuccessful in recolonizing its former area, especially in Europe. Another view still is that the plant survived the last, or even earlier Glacial periods, in ice-free coastal pockets within its present range but which are now submerged. There may have been a greater number of these refuges in America resulting in the present lop-sided distribution but this is speculation.

In the shallow water of loch margins, *Eriocaulon aquaticum* seems mainly limited in its occurrence by the effects of wave action. After winter storms rafts of plants may sometimes be seen strewn along loch shores. The growth and stucture of the plant must therefore maximise anchorage, strength and flexibility. From the moment of germination, the plant must be firmly secured and even before the seedling root (radicle) has emerged, the embryo of the plant develops into a hook-like structure which provides an initial anchor. The white, worm-like roots of the mature plant are highly specialized. Internal diaphragms provide strength, and soft tissue between them imparts pliancy. Similar diaphragms occur in the leaves and flowering stems. The characteristic jointed structure of the plant gives rise to the common name of pipewort. Every part of the plant is provided with air spaces for efficient gas diffusion. The flowering stem and leaves are provided with air canals but a more elaborate system occurs in the rhizome and roots. Here, cells in the cortex interconnect by spine-like projections and form a network of intercellular spaces.

The flowers, though minute, are structurally fascinating and beautiful. They are arranged in tight heads (or capitula) and the positioning of the separate male and female flowers is very variable. To promote cross-pollination the sexes mature asynchronously and occasionally totally unisexual heads occur. Both male and female flowers have a pair of sepals and petals and towards the tips of the latter there is a single ovoid nectary. This is larger in male flowers and to further augment the floral reward for insect visitors there is a pair of nectar-secreting glands in the centre of the flower. Because these glands are interpreted as a modification of the female parts (gynaecium) this is taken as evidence that the unisexual flower of *Eriocaulon aquaticum* has developed from ancestral hermaphrodite flowers.

Gentiana nivalis L.
ALPINE GENTIAN

This diminutive gentian is restricted to very few localities within the botanically rich mountain ranges of Ben Lawers and Caenlochan. The small size of the flowers is compensated by their intensity of colour and Reginald Farrer rightly remarked that the plant has "... eyes of a blue so violent that they even atone for the minuteness of the eyes themselves".

James Dickson, the Covent Garden nurseryman and founder member of the Linnean Society, first discovered the plant on Ben Lawers on a visit to the Highlands in 1792. It evaded detection in the Caenlochan district for another forty years until it was found in Glen Isla by Robert Greville. Although originally described as abundant at this site, the plant is now quite localized and is really only associated with Ben Lawers where some relatively large populations occur. Even on this mountain, despite careful

Gentiana nivalis, Perthshire, 1 viii 1995.

searches, some populations known in the 19th century have not been relocated in recent years.

Gentiana nivalis grows on south-facing ungrazed ledges and the grazed slopes below them. In these habitats, the plant is a rare associate of a type of high altitude calcareous herb-rich vegetation where a glittering array of other local and rare mountain plants occur such as alpine forget-me-not (*Myosotis alpestris*), alpine mouse-ear (*Cerastium alpinum*), cyphel (*Minuartia sedoides*), purple saxifrage (*Saxifraga oppositifolia*), rock whitlowgrass (*Draba norvegica*), alpine cinquefoil (*Potentilla crantzii*) and alpine saxifrage (*Saxifraga nivalis*). The flowering plants are equally matched in interest by the range of montane mosses, liverworts and lichens.

In Scandinavia and Central Europe *Gentiana nivalis* is less resticted to mountains. It is especially widespread in Iceland where it occurs on grass moors and heaths, flat boulder-strewn barrens (fellfields), low willow-scrub, open birchwoods and, most interestingly, in 'home fields' which is the manured grassland around farmsteads. The plant is well scattered in the European mountains reaching an altitude of 3100m in the Alps and extends eastwards to the Caucasus range. On the other side of the Atlantic, *G. nivalis* is restricted to the Labrador coast of east North America.

Long term population studies of alpine gentian on Ben Lawers by Scottish Natural Heritage and the Institute of Terrestrial Ecology have significantly increased ecological knowledge of the plant. With Iceland purslane (*Koenigia islandica*), *Gentiana nivalis* is one of the few montane plants that are generally considered as annuals. Strictly, annual plants complete their life cycle within twelve months, although this period may begin in the growing season of one year and finish in the next. Seed of alpine gentian may germinate either in late summer and autumn or in spring. Those that overwinter as young plants may survive beyond twelve months and are therefore, strictly, biennial. These plants tend to be larger, produce a greater number of flowers and a greater quantity of seed than spring-germinating plants. However, these more conspicuous plants appear to be preferentially grazed by sheep and if eaten before seed-set they will not contribute to the population. If ripe capsules are eaten seed may pass through the animals and dispersal can be facilitated. The occurrence of the plant in Icelandic homefields is interesting in this respect as seed may be transferred about the fields by animals. In Sweden the seed of *Gentiana nivalis* is known to be spread by cattle.

Like many annual plants, alpine gentian requires open patches in vegetation for seed germination and establishment of young plants. On Ben Lawers, in experimental enclosures protected from grazing and disturbance, *G. nivalis* gradually declined as it was outcompeted mainly by the spread of more vigorous perennial grasses. In unenclosed areas large, conspicuous plants of alpine gentian tended to be eaten but no general decline took place because breaks in the turf caused by trampling provided seed germination sites and therefore allowed another generation of plants to establish. However, this does not mean that sheep grazing is beneficial to *Gentiana nivalis* if adequate open sites in the turf form naturally within the range of seed dispersal. The soft and unstable Ben Lawers schist is easily eroded, rockfalls are common and bare ground is frequently formed. Buried seeds of alpine gentian can remain viable for several years so the plant is not entirely dependent on the previous season's seed production for continued survival. Like many annuals, its seed readily germinates in newly created sites. The seed-bank in fact allows for considerable seasonal variation in seed-set.

The main flowering period starts at the beginning of August but the flowers only open in bright sunshine. On dull days, when the petals are rolled together in spiral fashion, to protect the pollen from rain, the flowers are easily overlooked. The flower closing mechanism is very sensitive and on days with patchy cloud the flowers may open and close several times within an hour. Both cross- and self-pollination take place in alpine gentian. However, the flowers are adapted to cross-pollination as nectar is produced at the base of the ovary and the stigma projects beyond the anthers. Only insects with long tongues are able to reach the nectar at the base of the tubular flowers.

Goodyera repens (L) R. Br.
CREEPING LADY'S-TRESSES

Of all British orchids creeping lady's-tresses is the species that is most strongly associated with pinewoods. Other orchids such as lesser twayblade (*Listera cordata*), heath spotted orchid (*Dactylorhiza maculata*) and coralroot orchid (*Corallorhiza trifida*) may sometimes be seen within pinewoods, but grow mainly in other habitats.

James Robertson, a gardener at the Royal Botanic Garden Edinburgh and a pupil of the first Regius Keeper, John Hope, discovered *Goodyera repens* in 1767 "In a wood called Cragenon between the bridge of Nairn and Dalmagary". This was five years before John Lightfoot, who is generally credited with its discovery from the record in his *Flora Scotica* of 1777 where he states that he found the plant at Dundonnell, West Ross in 1772.

Although beaten by Robertson's record, Lightfoot's discovery is more remarkable because Dundonnell approaches the most westerly site known for *Goodyera* in Scotland and there would have been far less chance of coming across the plant there than around Nairn where Robertson discovered it.

Goodyera is chiefly a plant of the central and eastern Highland pinewoods with only scattered localities in the west although it reaches Arisaig and the area around Lochalsh. It is absent from most islands but has been recorded as far north as Orkney. The most southerly native locality was Houghton Woods near Hull where it persisted at least until 1888. Populations in Norfolk and Suffolk are thought to have arisen from introduction with pine seedlings from Scotland when plantations were established. In its main area of distribution *Goodyera* is a characteristic species of the drier pinewoods, the ground vegetation of which is dominated by dwarf shrubs such as heather (*Calluna vulgaris*) and bell heather (*Erica cinerea*) but often with abundant wavy hair-grass (*Deschampsia flexuosa*) and various common mosses. The scarce twinflower (*Linnaea borealis*) is an associate of *Goodyera* in some woods and the rarer one-flowered wintergreen (*Moneses uniflora*) is accompanied by *G. repens* at most of its sites. In a few localities, usually near the coast, *Goodyera* grows beneath heather outside the pine canopy.

Goodyera repens, West Ross, 25 vii 1995.

Goodyera repens has a virtually circumboreal distribution, and is nearly always found in mossy forests whether coniferous or mixed. It reaches approximately 70°N in Scandinavia and extends south to the Pyrenees. Eastwards it ranges through Europe and Asia to Kamchatka and Japan. It also reaches the Himalayas and temperate China. In the south of its geographic range *G. repens* becomes a montane plant although still confined to forest. The Scottish variety (strictly *Goodyera repens* var. *repens*) is rarer in North America and generally restricted to the Rocky Mountains and Canada whilst *Goodyera repens* var. *ophioides* is more widespread and grows in a range of coniferous woods and also in bogs. The varietal name *ophioides* means 'snake-like' from the distinctive white veining on the leaves. Variety *repens* lacks the white veins but the green marbling on the leaves can be both variable and beautiful; in some Scottish populations plants show a range of attractive leaf-patterning.

Goodyera is well adapted to nutrient-poor, shaded pinewoods. It is one of the few wintergreen orchids in Britain so can make use of the restricted light for photosynthesis all year round. Like all British orchids *Goodyera* has a mycorrhizal association and it has been demonstrated that this is essential for both germination and for healthy growth of the adult plant. Unlike true lady's-tresses which belong to the genus *Spiranthes* and have swollen tuberous roots, *Goodyera* has a creeping underground rhizome which produces slender branches that spread through the surface litter and humus layers of the forest floor. These branches produce rosettes of leaves at their tips which are able to act as individuals and produce flowers after about eight years. By this method of continuous vegetative multiplication the plant may achieve virtual immortality. However, to colonize entirely new areas and to remain genetically variable, seed must be produced. Although the leaf rosettes are able to tolerate extreme shade, *Goodyera* generally flowers best in dappled shade and the most spectacular displays are often seen where forest plantations have undergone thinning or where natural gaps have been created by windthrow. It is one of the later flowering British orchids, and the sweetly-scented flowers are mainly produced in July. Besides scent, nectar is also produced, both of which attract insect visitors which carry out cross-pollination. However, it has also been demonstrated that *Goodyera* is extremely self-fertile.

The genus *Goodyera* was named in honour of the 17th century English botanist John Goodyer (1592–1664) who recorded a number of plants for the first time in Britain and gave great service to Thomas Johnson in the preparation of the much improved second edition of *Gerard's Herbal* in 1633. In Johnson's edition, the text for the creeping satyrion, what we now call marsh helleborine (*Epipactis palustris*), was attached to an illustration of what is clearly *Goodyera repens*. The plant was described as growing "... plentifully in Hampshire, within a mile of a market Towne called Petersfield, in a moist medow named Wood-mead, neere the path leading from Petersfield, towards Beryton". This mistake went unnoticed and Robert Brown, in wishing to name a plant in honour of John Goodyer, chose one that grew near to where the latter resided. This unfortunately resulted in Goodyer's name being linked to a northern plant with which he would never have had any connection.

Koenigia islandica L.

ICELAND PURSLANE

What *Koenigia islandica* lacks in aesthetic qualities it makes up for in historical and ecological interest. This diminutive plant, named by Linnaeus in honour of one of his botanical pupils, John Gerard Koenig (1728–1785), was one of four important additions to our mountain flora made in a remarkable series of discoveries in the early 1950s. It had actually been discovered much earlier, on 31 August 1934, near the summit of the Storr on Trotternish, Isle of Skye and a dried specimen was deposited in the herbarium of the Royal Botanic Gardens, Kew. At the time it was mistakenly identified as a *Peplis* in the Loosestrife family (Lythraceae) whereas *Koenigia* is related to the docks and sorrels, family Polygonaceae. The error came to light in 1950, when Mr B. L. Burtt was examining the Kew collections of *Peplis* while dealing with a routine request. So *Koenigia* was added to the British flora. In 1956 its Scottish range was extended to Mull, with the discovery of plants on Maol Mheadhonach on the Ardmeanach peninsula. It has since been found on a few other hills in the vicinity.

On the mountains of the Trotternish ridge *Koenigia* occurs in areas of open wet gravel, formed from the easily weathered basalt and which are kept moist by spring-fed rills and flushes. The plant generally avoids drought-prone southerly aspects, although it occurs on the south side of the summit of the Storr and on some south-facing slopes on Mull. The sparse associated flora consists mainly of common yellow-sedge (*Carex viridula* subsp. *oedocarpa*), tufted hair-grass (*Deschampsia cespitosa*), starry saxifrage (*Saxifraga stellaris*) and several mosses. On Mull hairy stonecrop (*Sedum villosum*) is an interesting additional species. On Faeröe, where *Koenigia* is abundant, *Sedum villosum* and Arctic mouse-ear (*Cerastium arcticum*) are common associates but *Cerastium arcticum* occurs in only one place on the Trotternish ridge and not at all with *Koenigia* on Mull. On the steeper east-facing screes of the Trotternish hills *Koenigia* is joined by plants such as cyphel (*Minuartia sedoides*), spiked wood-rush (*Luzula spicata*), northern rock-cress (*Arabis petraea*), heath pearlwort (*Sagina subulata*) and mountain sorrel (*Oxyria digyna*). Here, open conditions favouring the annual *Koenigia* are maintained by the instability of the slopes which is intensified to some extent by grazing. On the ridge similar conditions are maintained by water movement and frost action in severe winters.

Koenigia islandica, Skye.
(Photo: P Lusby).

It is easy to imagine yourself in the barren desolation of Faeröe as you reach the Trotternish ridge from the east-facing slopes below. These atmospheric places share a similar geology. The basalt rocks of Mull and Skye are all parts of the Tertiary Igneous Province – a vast volcanic region that was active during this period, some 40 million years ago. However, *Koenigia* is not restricted to basalt. The most peculiar records are from a saltmarsh on Disko Island off the west coast of Greenland and from the margins of hydrothermal streams in Iceland where the temperature of the water is 70°C! Curiously, *Koenigia islandica* not only occurs in subarctic and montane regions of the northern hemisphere but also at the tip of South America, since there is now general agreement that *K. fuegiana*, from Tierra del Fuego, is the same species.

Koenigia is remarkable in being one of the very few annual arctic plants; most are herbaceous perennials with large underground storage organs which enable plants to begin growth immediately conditions become favourable in spring. The obvious problem for an annual in the arctic is the extremely short growing season; moreover, *Koenigia* grows at latitudes of up to 82°N which is a more polar distribution than for any other annual. It has a wide temperature optimum for growth, but is sensitive to heat. Because it is so low growing (2–6cm), it is greatly influenced by the temperature of the soil surface. Where the ground is wet, temperature variations will be small and evaporation tends to keep down the temperature on sunny days. In dry areas, soil temperatures may rise above 45°C and kill the plants. A low resistance to water loss through the leaves of *Koenigia* is thought to be a cooling mechanism, but also renders the plant vulnerable to water shortage. Furthermore, the very small leaves of *Koenigia* are advantageous in avoiding excessive heat.

Koenigia has a much more restricted distribution in Britain than formerly. Pollen of *Koenigia* has been identified from Late-glacial deposits in the south of Scotland, Cumbria, the Isle of Man and Northern Ireland. Like other arctic-alpines, it has had to retreat to areas with sufficiently cool summers where conditions are too severe for less well-adapted plants. *Koenigia* flowers from late June to August and towards the end of the season the leaves turn an attractive red.

Ligusticum scoticum L.

SCOTS LOVAGE

Robert Sibbald first recorded this attractive umbellifer in Britain in 1684, linking the plant with Scotland in his name *Imperatoriae affinis umbellifera maritima scotica* [the Scottish seaside umbellifer akin to *Imperatoria*]. This name is particularly apt as at the time he could not have known that besides its occurrence on the north and east coasts of Ireland, it is virtually confined in the British Isles to Scotland. Four years later John Ray was the first botanist to record a more exact locality for this Scottish speciality "on a certain sandy and stony hill six miles from Edinburgh towards Queensferry in Scotland".

Ligusticum scoticum,
Sutherland, 7 vii 1995.

Ligusticum scoticum may be found right around the Scottish coastline wherever suitable habitats occur. It is well represented on the Inner and Outer Hebrides, and even extends to St Kilda and the Northern Isles. The leathery leaves which are divided into three-stalked leaflets, and in turn divided into three rhomboidal-shaped lobes (biternate), are sufficient to distinguish the plant from other British umbellifers. The geographical distribution of *Ligusticum scoticum* is largely limited by its preference for cool summer temperatures, as it does not occur in areas where the mean July temperature exceeds 16°C. In its southern Scottish

localities it appears to grow better on cooler north-facing shorelines and at its southernmost limit in Britain, on the Northumberland coast, it is restricted to northerly aspects.

Scots lovage is distributed around the coastlines of countries on both sides of the Atlantic. In Europe it occurs on rocky seashores from Denmark to north Norway (including Faeröe) and further east it can be found in north Russia along the coast of the Barents Sea. *L. scoticum* extends across Iceland and west Greenland to the east coast of North America and reaches south as far as the southern end of Hudson Bay. On the Pacific side of the globe the plant is represented by subspecies *hultenii*, which occurs in Alaska, Siberia, northern Japan and Korea. The main differences between subsp. *hultenii* and subsp. *scoticum* are that the former has larger fruit, smaller flower heads and leaf veins that form loops at the leaf margin rather than having blind ends.

Ligusticum scoticum is sensitive to physical damage and does not occur in areas that are grazed or trampled. It is therefore generally confined to cliff crevices and ledges that are inaccessible to herbivores and man. It also grows more vigorously in relatively sheltered sites where it escapes the full force of salt-laden gales. This appears to contradict its Gaelic name of *siunas* from *sion* meaning 'a blast' or 'storm' presumably derived from the exposed coastal rocks on which the plant can survive. It is not clear whether the absence of the plant from seabird cliffs is due to raised nutrient levels or to physical damage by birds.

In its typical habitat *L. scoticum* is a distinctive member of a particular maritime cliff community and is usually accompanied by thrift (*Armeria maritima*) and red fescue (*Festuca rubra*), with frequent sea campion (*Silene uniflora*) and roseroot (*Sedum rosea*). On cliff tops where the crevice and ledge community gives way to more continuous vegetation, *L. scoticum* may accompany a wider range of grassland plants including bird's-foot trefoil (*Lotus corniculatus*), white clover (*Trifolium repens*), Yorkshire-fog (*Holcus lanatus*) and autumn hawkbit (*Leontodon autumnale*). It does not thrive in grassland communities, as the leaves tend to rot in wet conditions and the seeds require more open habitats for germination. In similar coastal habitats in England and Wales *L. scoticum* is replaced by what may be thought of as its southern counterpart, rock samphire (*Crithmum maritimum*). The British distribution of these two species is almost mutually exclusive, apart from a slight overlap mainly in Kirkcudbrightshire and Wigtownshire.

Aspects of the biology of *L. scoticum* show the plant's obvious adaptation to the northern maritime environment. The plant is more frost tolerant than *Crithmum maritimum* which is probably the major factor accounting for the replacement of the *Crithmum* ledge community by the *Ligusticum scoticum* community north of the Mull of Galloway. Although Scots lovage seems to positively benefit from moderate exposure, seed production is higher in sheltered sites, possibly because pollinating insects are more abundant. The fruit is buoyant in salt water and is able to float for two to three months, even retaining some viability after a year at sea. A cool wet period is required for germination and although *L. scoticum* can grow in stabilized shingle, or occasionally on fixed dunes, it will not grow where the ground is unstable.

Suitable conditions for germination are important both in the short and long term because the plant reproduces entirely by seed with no effective vegetative reproduction. Like many northern and arctic plants, *L. scoticum* is adapted to the short growing season, and with a high rate of respiration it can rapidly start into growth when the temperature begins to rise. In favourable conditions the leaves can reach their full size within two to three weeks, but in unusually warm conditions the plant could suffer by growing too quickly and depleting its food reserves. In contrast, *Crithmum maritimum* has been shown to have a much slower respiration rate and therefore grows more slowly in the warmer southern climate.

Ligusticum scoticum flowers from late June to August with the seeds ripening in October and November. The leaves have formerly been cooked as a vegetable and sometimes eaten as a preventative against scurvy. Its taste is apparently similar to celery but the 18th century botanist Sir James Edward Smith maintained that "The flavour is highly acrid, and though aromatic, and perhaps not unwholesome, very nauseous to those who are unaccustomed to such food".

Linnaea borealis L.

TWIN FLOWER

Scottish pinewoods are of great natural beauty and conservation significance but are not rich habitats for flowering plants. However, the three rarities normally associated with them, *Linnaea borealis*, *Moneses uniflora* and *Goodyera repens*, are among the most beautiful Scottish wild flowers. *Linnaea* is particularly delicate. Named by Jan Frederik Gronovius in honour, but at the behest of Carl Linnaeus (1707–1778), this celebrated naturalist chose well. His well known description "... a plant of Lapland, lowly, insignificant, disregarded, flowering but for a brief space – from Linnaeus who

Linnaea borealis, Sutherland.
(Photo: P Lusby).

resembles it", is only meant in mock modesty but is hardly true since the sight of a carpet of profusely flowering *Linnaea*, viewed in slanting light under a pinewood canopy is far from lowly and insignificant and the beauty, on closer scrutiny, is hard to equal.

The plant was first recorded in Britain by James Beattie (1735–1810), Professor of Natural History at Aberdeen. He found it in a pinewood at Inglismaldie, Mearns, near Aberdeen in 1795. Further localities were soon discovered and by 1855 William MacGillivray in his *Natural History of Dee Side and Braemar*, mentioned a further seven

Linnaea borealis, Sutherland,
21 vi 1995.

sites. By 1821 *Linnaea* had been recorded south of the Border from Catcherside, Northumberland but is doubtfully native there as the conifers under which it grows were imported from Norway, where the plant is common, in about 1770. There are also records from north-east Yorkshire and Durham.

Linnaea has a predominantly north-eastern distribution in Scotland (although it has been recorded from near Fort William) and grows mainly in drier pinewoods in association with other dwarf shrubs such as heather (*Calluna vulgaris*) and blaeberry (*Vaccinium myrtillus*), with wavy hair-grass (*Deschampsia flexuosa*) and several dominant mosses, the commonest of which is usually *Hylocomium splendens*. Where other dwarf shrubs are vigorous *Linnaea* may be suppressed. It occasionally occurs in birch woods and in a few Scottish localities the plant grows outside woodland in open herb-rich heath, as in the Morven area of Aberdeenshire, and also in the shade of rocks. In Greenland the plant occurs far more abundantly in open heathland. Plants in woodland and heathland differ in the structure and size of the leaves as well as growth form.

Like many of our rarities, the occurrence of *Linnaea borealis* in Scotland represents a north-west Atlantic extension of its Continental distribution. As a species, *L. borealis* has a circumboreal distribution but the Scottish plant (strictly *L. borealis* subsp. *borealis*) meets the slightly rounder-leaved and more tubular-flowered *L. borealis* subsp. *americana* in Alaska and is replaced by it in North America. Further along the Pacific Coast of America the more acute-leaved, long-flowered and rather woody-stemmed subspecies *longiflora* occurs. Although only three subspecies are recognized in this very widely distributed species, the Swedish botanist, Veit Brecher Wittrock, published an exhaustive study of the variation of *Linnaea borealis* subsp. *borealis* in 1907. He paid particular attention to the colour and patterning of the corolla and named over 150 forms chiefly based on these characters. This was an interesting exercise but this is taxonomic classification for the specialist and very few botanists uphold Wittrock's names. No doubt variants within our Scottish populations would be referable to some of his forms.

Like many plants of the forest ground layer, *Linnaea* forms large clonal patches by vegetative spread. The main shoots that grow over the surface (stolons) bear either lateral, non-flowering leafy shoots or flowering shoots. By disintegration of older stems separate patches are formed. The main flowering period is June and the nectar-rich flowers are visited by short-tongued insects. Flies, solitary bees and hover-flies are the most common pollinators. In Scottish populations flies seem to be the most important although their visits to flowers are rather infrequent. A further problem is that there is a reduced level of fertility between plants of the same genotype so, where a population consists of very few clones, seed-set is low. Clonal plants mainly increase vegetatively but are dependent on stable conditions and are often slow to colonize new sites. The presence of large patches of *Linnaea* is a useful indicator of ancient or long-established habitats. However, *Linnaea* is remarkably mobile on the forest floor as the stolons can grow rapidly in favourable conditions (up to 48cm per year has been recorded) and can grow over obstacles. In this way the plant is less constrained by rocks beneath the soil surface than species with rhizomes. The one-seeded fruits are adapted for animal dispersal by being partially enclosed by small, glandular hairy bracts. When touched by a passing animal the fruit stalk breaks below the bracts and the fruit sticks to either fur or feathers.

The rarity of *Linnaea* in Scotland has prevented the plant from becoming familiar to many people, hence there are very few vernacular names, commonly twinflower, linnaea or two-flowered linnaea. However, in Scandinavia, where the plant is more widespread it has names such as wood queen, perfume flower, little lily-of-the-valley, and bridal veil.

Loiseleuria procumbens (L.) Desv.

TRAILING AZALEA

James Robertson found *Loiseleuria procumbens* on several mountains, including Ben Horn, Beinn Mhealich, Scaraben and Ben Wyvis during his tour through the north of Scotland in 1767. He also found it on Ben Avon in the Cairngorms in 1771, where he was struck by the barrenness of the granite tops which he said "... produce little but the procumbent Azalea ...".

Loiseleuria procumbens is a true alpine whose British distribution reflects that of land over 650m, predominantly in the Cairngorms and the north-west Highlands. Southwards it reaches the mountains around Loch Lomond, and northwards Orkney and Shetland, where it descends to its lowest altitude of about 400m. Fossil remains of the plant, which have been dated to the Late-glacial, around 12,000 years ago, when the summer temperature is reckoned to have been at least five degrees lower than present, have been found at Corstorphine in Edinburgh and in north Wales.

In northern Europe *Loiseleuria* is common throughout the Scandinavian mountains and extends eastwards across the Asian tundra to northern Japan. It occurs in all the main mountain ranges of central Europe, Iceland, Faeröe, Greenland, Alaska and Newfoundland, with its southernmost outpost in the mountains of New Hampshire. The total range of the plant is arctic-montane, tending to avoid only the more extreme continental regions where the temperature falls below -40°C.

Loiseleuria procumbens,
Inverness-shire, 7 vi 1995.

Trailing azalea is typically a plant of exposed montane heaths where the snow cover is thin and of short duration, and even survives where the wind speed at ground level can average two to three metres per second. Its small leaves have very thick cuticles which reduce water loss and protect the inner tissues against abrasion from blowing ice crystals. The stomata are very sensitive to wind and close quickly in response to increased wind velocity. By this means *Loiseleuria* is better able to reduce its transpiration than other dwarf shrubs and can therefore grow successfully in more extreme habitats. Whilst preferring snow-free, windswept sites, during times of alternate freeze and thaw when little water is available in the shallow soil, *Loiseleuria* is dependent on the moisture from melting snow or ice. The individual leaves are able to take up water rapidly through two hairy, thinly-cutinised channels on the undersides which act like capillaries running the length of the leaf. In warmer periods dew, which is frequent where the daily temperature range is large, is also important for the plant's survival. The plant also absorbs water through numerous adventitious roots borne on the stems. Dense mats of *Loiseleuria* filled with leaf litter can store a large amount of water and are functionally similar to a large succulent leaf. Where branches grow out over rocks they often show drought damage, whereas branches lying over humus remain unaffected.

Although less frequent than extreme cold, *Loiseleuria* has also to contend with extreme heat. The sun can warm up the ground considerably in some exposed habitats, even in winter. A leaf temperature of 43°C has been recorded in the Alps for *Loiseleuria*, when the surrounding air was only 23°C. It has been determined that the plant has a maximum lethal temperature of 51.8°C and, although this may be rarely reached, heat damage is possible on clear sunny days when the wind drops briefly. Moderate rises in leaf temperature may benefit the plant when the ambient air is too cold to permit photosynthesis, but as the leaf temperature increases so does the respiration rate and, in such situations the plant is in danger of losing a large amount of energy. *Loiseleuria* is thought able to cope with this since it stores large amounts of fat in its tissues.

The plant's most abundant associates are lichens which often dominate the ground cover. Of the other dwarf shrubs in this windswept vegetation, *Arctostaphylos alpinus* is the most characteristic, both *Arctostaphylos* and *Loiseleuria* being 'espalier heaths', whose prostrate spreading branches hug the ground closely as if literally tied down.

The *Loiseleuria*-lichen heath or 'wind-carpet' is a specialized community that generally forms quite small patches which are determined largely by topography. Even the shelter of a large stone can favour the growth of a grass-sedge sward or larger dwarf shrubs such as juniper (*Juniperus communis* subsp. *alpina*) and bilberry (*Vaccinium myrtillus*). On mountains the community disappears in the upper montane zone as snow cover is greater and shelter from the wind increases. Furthermore, both *Loiseleuria* and the surface-encrusting lichens can only spread on stable ground which is not constantly disturbed by erosion or solifluction. Since it grows so very slowly, *Loiseleuria* requires a long time to cover the ground and form the small-scale mosaic characteristic of the community. By examining the annual rings of the woody central stem, both the annual growth rate and the age of the plant can be determined. These rings average less than one tenth of a millimetre thick, and a 64 year old plant has been recorded from Russian Lapland. If undisturbed, *Loiseleuria* can undoubtedly live longer than this. In some of its Scottish sites it has persisted for at least two centuries.

The genus *Loiseleuria*, which contains just the one species, was named after the prominent French botanist and physician Jean Louis Auguste Loiseleur-Deslongchamps (1774–1849). The plant has been called the world's smallest azalea and its affinities to the *Rhododendron-Azalea* group are obvious. *Loiseleuira* differs primarily in having opposite rather than alternate leaves. Its delightful red buds, like little plums on the end of the shoots, open between May and July and the small starry flowers which are pink at first fade to almost white.

Lychnis alpina L.
ALPINE CATCHFLY

Lychnis alpina was first found in Britain in 1795 by George Don on Little Kilrannoch at the head of Glen Doll in Angus. The exact whereabouts of the site remained a secret for several years, prompting one botanist to accuse Don of sowing the plant from seed. However, the secret did not last and in the 19th century *L. alpina* was heavily collected by botanists, doubtless attracted by the plant's rarity and charm. For many, a specimen of such a beautiful plant served as a just reward for the physical effort expended in reaching the plateau.

L. alpina survives in just one other site in Britain, Hobcarton Crag, near Keswick in Cumbria. The species formerly occurred on Old Man of Coniston in north Lancashire but disappeared from there sometime before 1950. There is also an old record from Rhum, dated 1942, but the plant has never been seen there since.

A major decline in the distribution of *L. alpina* has occurred since the end of the last glaciation, as fossils of the plant dated to this period have been found as far south as Essex and on the Isle of Man. As the climate warmed and forest encroached on formerly open lowland habitats, *L. alpina* would have become progressively restricted to high ground and sites otherwise uncolonized by trees. The development of peat during the Atlantic period would have eliminated the species from many areas, including limestone districts. Although such substrates may have been later re-exposed, the distribution of the species had become so reduced and fragmented that it failed to recolonize.

Lychnis alpina is an example of an Amphi-Atlantic species, with a distribution split either side of the north Atlantic. Populations in eastern North America and Greenland are recognized as subspecies *americana*, whilst European populations are referable to either subspecies *borealis* or *alpina*. The former describes plants in Iceland and on ultrabasic rocks in Scandinavia, and the latter the British and other European populations, including those in the Alps, Pyrenees and Caucasus mountains. *L. alpina* is best described as a subarctic-alpine, with scattered occurrences in the lowlands.

Lychnis alpina, Angus, 9 vii 1995.

Little Kilrannoch is the most montane occurrence of serpentinite in Britain. Here *L. alpina* grows between 850 and 870 metres on outcrops of serpentinite debris which interrupt an eroding peat surface. Serpentinite soils are inimical to the greater number of plant species. The debris on Little Kilrannoch supports a sparse but unique flora containing thrift (*Armeria maritima*), cyphel (*Minuartia sedoides*), mountain scurvygrass (*Cochleaia micacea*) and a distinctive form of common mouse-ear (*Cerastium fontanum* subsp. *scotica*) found only on this outcrop. Among the unfavourable chemical properties of the soil are a high magnesium to calcium ratio, high concentrations of the heavy metals nickel and chromium, and very low amounts of major nutrients. It is also physically unstable, a condition due to the peculiar sliding propensity of weathered serpentinite which is often described as a 'soapy' rock, and exacerbated by the strong influence of frost heave in this exposed montane site. Special soil conditions also characterize Hobcarton Crag where *L. alpina* grows at around 600m on an outcropping quartz vein associated with pyrites. Among its associates here is the rare copper moss, *Grimmia atrata*, which is otherwise found on old copper mines.

Because many factors contribute to the unfavourable conditions for plant growth on serpentinite, there appears no unifying adaptation that allows a particular species to survive on it. One important characteristic of *L. alpina* is its ability to store magnesium, as crystals of magnesium oxalate, inside its cell vacuoles, and therefore avoid the potential metabolic imbalance due to a high ratio of magnesium to calcium.

The majority of the Pink family (Caryophyllaceae) are so-called oxalate plants which may explain why it is so well represented on serpentinite. With the exception of thrift, the most frequent flowering plants on Little Kilrannoch serpentinite belong to this family.

L. alpina is also able to accumulate heavy metals from soils which are regarded as toxic to the majority of plants. In Sweden it occurs widely on serpentinite, but also very commonly on ore and slag deposits, especially on old mine sites with mineral soils rich in copper, zinc or lead. Such is the plant's association with heavy metal soils that it has been used as a geobotanical indicator of ore deposits. Scandinavian pioneers of geobotany in the 17th century called *L. alpina*, 'kisplant', pyrite plant, or 'kobberblomst', copper flower.

Alpine catchfly lacks the sticky stems of the sticky catchfly, *Lychnis viscaria*, and is altogether smaller and more compact in habit. Older plants tend to form small mounds of dense leaf rosettes, resembling the typically mounded cushions of thrift (*Armeria maritima*). These are raised above the debris surface by frost-heave and surface erosion, but the plant's strong central tap root provides sufficient anchorage against up-rooting. The leaves die down in winter (or in spring if snow-covered) leaving only the tap root remaining alive with next season's leaf buds sheltering under the persistent withered leaves. Although individual plants may live for many years, *L. alpina* is ultimately dependent on seed-set for survival of its populations as the plant has no means of vegetative propagation or dispersal.

Lychnis viscaria L.
STICKY CATCHFLY

There are few nationally rare plants that have managed to survive for over 300 years in a city centre but *Lychnis viscaria* is the rare jewel of Edinburgh. It was first reported as native in Britain by the country's finest 17th century botanist, John Ray in his *Catalogus Plantarum Angliae* of 1670 "... *In Rupibus in vivario Edinburgensi ...*" [on rocks in Edinburgh Park] but it was Thomas Willisel, employed as official collector to the Royal Society of London in 1668, who actually discovered it there. Although *L. viscaria* still survives at Willisel's original site on Samson's Ribs, fires in recent years have reduced the population to a few inaccessible clumps, leaving the species at this historically important site in considerable danger of extinction. All other known populations in Edinburgh have been lost through collecting, rockfalls and quarrying.

In Scotland (and at two Welsh sites) *L. viscaria* is strongly associated with basic and intermediate igneous rocks and a warm, dry climate. Accordingly it has an eastern tendency and usually occupies crevices and ledges on south- to south-east-facing rocks. In Europe it grows on a wider range of rocks including sandstone and limestone. The plant is much scarcer in Scotland on metamorphic and sedimentary rocks with the notable exception of the flourishing coastal colony in Kirkcudbrightshire which grows on unstable sedimentary mudstone undercliff.

Lychnis viscaria at its most westerly Scottish site, Kirkcudbrightshire, 17 v 1995.

The main concentration of *L. viscaria* in Britain is in the western Ochil Hills where it occurs mainly on precipitous cliffs of andesite. It also extends into Fife on the eastern fringe of these hills. Although predominently a plant of rock outcrops, it is not montane (unlike the rarer *L. alpina*) and only reaches an altitude of about 425m.

The juxtaposition of plants that grow on acid and basic rocks are very much a feature of the associated vegetation of *L. viscaria* and it is not unusual to see the base-loving meadow oat-grass (*Helictotrichon pratense*) for example, growing in close proximity to heather (*Calluna vulgaris*), bell heather (*Erica cinerea*), gorse (*Ulex europaeus*) and broom (*Cytisus scoparius*). This phenomenon probably occurs because the base status of the soils lies within a fairly narrow range (slightly acid) where most plant nutrients are available. Outwith this range some nutrients may become limiting and others toxic to calcicoles and calcifuges. Furthermore, the dryness and shallow depth of the soil keep nutrient levels below that allowing the establishment of aggressive plants of more fertile soils.

L. viscaria is both intolerant of shade and extremely vulnerable to grazing animals. At some of its locations, in Glen Farg, Perthshire for example, the plant grows on less steep slopes and has been threatened by developing woodland. In shaded conditions *L. viscaria* fails to flower and set seed. This may not affect populations in the short term as individual plants live for several years; but as the plant relies on seed for reproduction and dispersal, long term affects are more serious. In very shaded conditions even vegetative plants become weak and die. Although the flowers are self-compatible, cross-pollination results in a greater number of viable seed. The latter is promoted because pollen is shed within individual flowers before the stigmas are receptive (protandry). Butterflies, bees and hover-flies are the main pollinators and only those with long tongues can reach the nectar at the base of the corolla tube. However, some short-tongued insects steal the nectar by biting through the base of the flower. These so-called 'nectar robbers' do not collect pollen and reduce the availability of nectar for legitimate pollinators. Bees are the most efficient pollinators within a *Lychnis* colony, but butterflies are potential long distance vectors, which is important for genetic interchange between populations. It is essential that colonies of *L. viscaria* do not become too fragmented, as isolated populations suffer reduced visits by pollinating insects.

The brilliant red-purple flowers well earn the Greek generic name *Lychnis* [a lamp] when they light up their rock ledges and crevices from mid-May to mid-June but they may be harbouring a more iniquitous foe than a nectar-robbing insect. This is an anther smut fungus that replaces the pollen with fungal spores and also prevents ovule formation. The flowers are rendered entirely sterile and, moreover, the longevity of infected plants is not altered so they are transformed into disease reservoirs for their natural life span.

Lychnis viscaria growing on sedimentary undercliff, Kirkcudbrightshire, 17 v 1995.

Lychnis viscaria is unlikely to be confused with many other plants. The flower colour is similar but brighter than red campion (*Silene dioica*) and plants viewed from a distance can be difficult to distinguish. However, the much narrower leaves of the *Lychnis* and the characteristic sticky or viscid (hence 'viscaria') secretion on the flowering stems readily identify it. The secretion is generally thought to be a defence against nectar and pollen robbing insects and since several small insects are caught (hence 'catchfly'), interest has been raised amongst carnivorous plant enthusiasts but insectivory has not yet been proven. The eminent botanist, Sir James Edward Smith, took the view that " ... their [insects] decaying bodies form an air salutary to vegetable life".

As *L. viscaria* is easy to raise from seed, it has become quite a popular garden plant and several horticultural selections have been made including a double-flowered form ('plena'), and a white-flowered form ('alba'). Wild plants of the latter were recorded from two former localities at Blackford Hill in Edinburgh and from the Den of Airlie in Angus by George Don, the famous Forfarshire botanist. Less striking colour variants and dwarf plants have been given horticultural names but the only other botanical subspecies, *atropurpurea*, is confined to Romania and the Balkan Peninsula. Strictly the British plants should be referred to as *Lychnis viscaria* subsp. *viscaria*.

Mertensia maritima (L.) S. F. Gray

OYSTER PLANT

This scarce and beautiful member of the borage family (Boraginaceae) is a plant one always hopes to find when botanizing on northern shingle shores. Its tendency to come and go at sites at the whim of winter storms makes searching for the plant almost as unpredictable as bird-spotting.

The discovery of *Mertensia maritima* in Britain was reported by John Parkinson in his herbal *Theatrum Botannicum* of 1640. It was found by a Mr Thomas Hesket on "... One of the Iles about Lancashire ..." [probably Walney] and, consequently, Parkinson coined the name of Lancashire Buglosse. By the time John Lightfoot's *Flora Scotica* was published in 1777 he knew *Mertensia* from several Scottish localities including Bute, Fife, Arran and Glen Elg in Inverness-shire, and regarded it as one of the most beautiful of British plants – "... its undulated glaucous leaves contrasted with red and blue flowers, are extremely ornamental to the barren shores where it grows, and readily discover the plant to any curious observer".

Mertensia maritima,
Wigtownshire, 16 v 1995.

The plant has been recorded from scattered localities round the coast of the British Isles but its present headquarters is in Orkney with smaller concentrations in the Inner Hebrides and along the Banffshire coast. In Ireland it is confined to the north and north-east coast, but there are old records extending south to Wicklow and north Kerry on the east and west coast respectively. On mainland Britain the most southerly records generally accepted are from north-east Norfolk and Cardiganshire. Those from Cornwall, Devon, Hampshire and Lincolnshire are less certain.

Mertensia maritima is strictly a plant of the seashore and is well adapted to growing on shingle. It tolerates being buried by pebbles and may even benefit from the added protection in severe storms. It also occurs on less coarse material but rarely anything finer than coarse sand. On shingle beaches it may form almost pure stands but the most common associates are sea purslane (*Honckenya peploides*), sea campion (*Silene uniflora*), common sorrel (*Rumex acetosa*), curled dock (*Rumex crispus*), cleavers (*Galium aparine*), Babington's orache (*Atriplex glabriuscula*), and silverweed (*Potentilla anserina*). Colonies of *Mertensia* are often strewn with seaweed which may provide a source of nutrients.

The global distribution of *Mertensia maritima* is extensive. It occurs on both sides of the Atlantic Ocean and reaches its northern limit in Spitzbergen. On the European side of the Atlantic its southern limit is in Britain but on the eastern and western coasts of North America it extends south to 42°N and 50°N respectively. The plant is absent from the cold Siberian coast to almost the Bering Strait probably because of the short time the shores remain ice-free. Around the Pacific coast of eastern Asia the larger-flowered subsp. *asiatica* replaces the widespread subsp. *maritima*, but in the Aleutian Islands the ranges of the two overlap and plants intermediate between them have been recorded.

Like Scots lovage (*Ligusticum scoticum*) and other northern species, *Mertensia maritima* is adapted to short growing seasons by respiring at high rates at low temperatures enabling it to grow quickly early in the season. In warmer climates this adaptation can lead to rapid loss of food reserves. Besides this feature of the plant's metabolism, cold winters stimulate seed germination. The southerly geographical range of the plant is considered to be mainly limited by warmer temperatures.

Seed production is vital for long-term survival as no natural method of vegetative propagation has been observed in *Mertensia*, although severed pieces of root establish easily in cultivation. The main flowering period is from June to August and self-pollination occurs regularly. Seeds are dispersed, at least to some extent, by the sea and remain afloat for several days without being adversely affected. Seedlings are well adapted to the unstable substrate and their powers of elongation can raise them to the surface from beneath some 10cm of shingle. Although populations of *M. maritima* are vulnerable to destruction by severe winter storms, the roots of mature plants form cable-like structures for maximum achorage. This is achieved by an unusual method of internal thickening in the tap root and subsequent division into separate strands. The largest and most stable populations of *Mertensia* are those which occur on beaches sheltered from the full force of the waves by off-shore rocks and skerries.

The general northward migration of *M. maritima* and consequent loss from southern sites in Britain may be related to changing climate but the main threat to populations still within their climatic tolerance is grazing animals. Besides the loss in seed production when flowering stems are grazed, shoots bruised by trampling die back to the base. Other threats include shingle extraction and other forms of human disturbance of the habitat. To provide the best chance of conserving the British population of *Mertensia* there must be adequate suitable sites for this mobile plant to move between. This poses an awkward problem because it is always more difficult to defend sites for their potential, rather than actual, conservation importance.

Minuartia sedoides (L.) Hiern.

CYPHEL

The common name of cyphel for *Minuartia sedoides* is derived from the Greek *kuphella* meaning 'hollows of the ear' and also 'clouds of mist'. In the absence of specific medicinal reference to the use of the plant for hearing defects, the name is most likely used in the latter sense as an indication of the plant's habitat, on mountains which are often enveloped in cloud.

The first definite Scottish record for the plant was another made by James Robertson on his pioneering botanical tour of the north of Scotland. In June 1767 he discovered the plant on Ben Klibreck near Loch Naver. "Here I found only one plant which I had not before seen on Benevis [Ben Wyvis]. This was *Cherleria* [*Minuartia*] *sedoides*/Dwarf Cherleria. It grows abundantly on the north side of this hill, but was never known before to be a British plant." The following month he found the plant again on Ben Griam to the east of Ben Klibreck. These records pre-date that of the generally accepted first record for the plant, reported in John Lightfoot's *Flora Scotica* of 1777, which is based on specimens he collected from Baikeval on the island of Rhum in 1772.

Minuartia sedoides is a scarce plant in Britain and is confined to Scotland, having never been recorded from the mountains of England and Wales. It has a scattered distribution with records from Argyll to Caithness on the mainland as well as the Inner Hebrides. It is normally found above 457m and reaches 1190m on Ben Lawers. However, it has been recorded at 213m as a washed down plant in the stony bed of the River Fillan near Tyndrum. Records based on herbarium specimens collected at sea level from the mouth of the North Esk in Angus appear to be mistakes: one specimen has been identified as a form of annual knawel (*Scleranthus annuus*) and a second plant is *M. sedoides* but is believed to have been collected from the mountains. Records from Shetland are almost certainly errors having arisen from confusion between *M. sedoides* and non-flowering moss campion (*Silene acaulis*). There are several genera with small, narrow (linear) leaves in the pink family (Caryophyllaceae) which occur in Scottish mountains. These are quite easily confused with each other when not in flower. However, *Silene acaulis* and *Minuartia sedoides* are the only two where the margins of the leaves are fringed with very small colourless teeth. The teeth (or cilia) of *Silene acaulis* are more pronounced than those of *M. sedoides* but, more importantly, the leaf

Minuartia sedoides,
Sutherland, 14 vi 1994.

apices of the latter are blunt whilst those of *S. acaulis* are acute and end in a sharp point.

Minuartia sedoides does not seem to have a strong preference for a particular rock type but is generally more abundant on basic rocks such as basalt, mica-schist and serpentinite. It occurs in the same dwarf-herb communities as *Silene acaulis* but it is slightly more restricted in that, unlike the *Silene*, it does not grow in snow beds, dripping banks or tall-herb vegetation on mountain ledges.

The disjunct world distribution of *Minuartia sedoides* is fascinating. Besides Scotland, the plant is confined to the Pyrenees, Alps, Carpathians and the mountains of Bosnia. It is one of the very few Scottish montane plants that fail to reach the Arctic and is completely absent from Scandinavia. It has a truly alpine distribution. A plant with a similar range is purple oxytropis (*Oxytropis halleri*) but its habitat is not so montane in Britain. The close proximity of *Minuartia sedoides* and the arctic-annual *Koenigia islandica* on the basalt screes and gravel pans of the Trotternish hills in northern Skye form a unique assemblage of great botanical interest.

The greenish-yellow flowers of *M. sedoides* generally lack petals, a feature of the genus *Cherleria* to which the plant formerly belonged. The plant is polygamous, having variable mixtures of functionally male, female and hermaphrodite flowers on the same or different plants. Individuals with mostly functionally male flowers are particularly attractive as the comparatively large pale yellow anthers contrast with the sepals and render the flowers surprisingly conspicuous. Rudimentary petals sometimes develop in male flowers but are very small. Pollinators are mainly small flies.

The abundance of *Minuartia sedoides* at many of its Scottish sites gives the impression of a plant of wide distribution. Only by considering the global distribution of the plant can one appreciate how internationally important the Scottish population is.

Moneses uniflora (L.) Gray
ONE-FLOWERED WINTERGREEN

Moneses uniflora is the rarest and arguably the most beautiful flowering plant of the Scottish pinewoods. *Linnaea borealis* is of similar delicacy, but there is a certain fragility about *Moneses* which is so appropriate since it is particularly vulnerable to disturbance.

The first record of *Moneses* as a Scottish native is highly doubtful: two specimens in the herbarium of Sir J. E. Smith labelled "From the Western Isles of Harris and Berneray gathered in 1783 by James Hoggan". The Western Isles have never been well-wooded, especially with pine. It seems impossible therefore that the plant could have survived. A more likely explanation is that the specimens were mislabelled! Moreover, the Outer Hebrides are a long way outside the geographical area of all authenticated records for the plant. The first definite discovery was in 1792 when James Brodie found *Moneses* in the pinewoods near Brodie House, Nairn. In the same year James Hoy, secretary and librarian to the Duke of Gordon at Gordon Castle, collected specimens locally. Sadly, the beauty of the plant led to its downfall. In a number of localities *Moneses* has been reduced to very small populations or made extinct through botanical collecting. The most notable loss is the once strong population at Scone in Perthshire where, from the first notice of its existence in 1825, it was gathered almost continuously until 1883. Particularly avid collectors were William Gardiner and Colonel Henry Maurice Drummond-Hay, the latter of whom attached 25 flowering plants to a single herbarium sheet! The last record from Scone is dated 1922. From a former distribution in 12 botanical vice-counties, *Moneses* is now known from only three: Moray, Easterness and East Sutherland.

Moneses uniflora, Sutherland, 22 vi 1995.

The plant's general distribution in Scotland is markedly north-eastern, centred in Moray but extending west to Strathfarrar and Glen Affric. *Moneses* grows most abundantly within old Scots pine plantations which have established over coastal sand and usually occurs with the ground vegetation dominated by a varied mosaic of heather (*Calluna vulgaris*), blaeberry (*Vaccinium myrtillus*) and wavy hair-grass (*Deschampsia flexuosa*) with associates such as crowberry (*Empetrum nigrum*), cowberry (*Vaccinium uva-ursi*), creeping lady's-tresses (*Goodyera repens*) and occasionally *Linnaea borealis* and lesser twayblade (*Listera cordata*). A carpet of mosses is virtually always present, the most common of which is *Hylocomium splendens*. At a single Scottish site near Elgin, *Moneses* grows in a wet grassy mire but does not appear very healthy. This area was formerly wooded.

Even though the plant has been significantly reduced in Scotland through collecting and changes in landuse (mainly felling of native pinewood), it may never have been as common here as in Scandinavia, where it is primarily a plant of damp spruce forests and occurs secondarily in drier pinewoods.

Moneses has a circumboreal distribution ranging through northern Europe, Asia and India to Japan. In North America it is widespread in Canada but more restricted to northern states of the USA although it reaches Mexico.

The fidelity of *Moneses* to coniferous woodland reflects a more exciting adaptation to the shaded, moist forest floor and nutrient-poor conditions than other close relatives such as serrated wintergreen (*Orthilia secunda*) and common wintergreen (*Pyrola minor*) which occur in a wider range of habitats.

The layer of mosses that cocoon the leaf rosettes of *Moneses* play an important role in moderating the microclimate around the plants. They retain moisture so humidity levels are maintained and temperatures are also elevated immediately above the forest floor. The structure of the leaves of *Moneses* renders the plant less able than its relatives to withstand dryness so it is more vulnerable to fluctuating moisture conditions. In winter the moss layer provides protection for its buds and also reduces the depth to which frost can penetrate the soil.

The creeping, thread-like root system of *Moneses* spreads for considerable distances at the base of the humus layer, and produces daughter plants from root buds. In this way large clonal patches are formed. The roots are heavily infected with mycorrhizal fungi that supply essential nutrients to the plant. Other members of the Pyrolaceae, the small family to which *Moneses* belongs, are not so dependent on mycorrhizae and possess food-storing rhizomes. They are therefore less tied to the humid forest floor where the fungal association is of greatest importance.

The beautiful solitary white flower of *Moneses* is produced in June and was described by Reginald Farrer as having "a scent of orange blossom, so poignant in its deliciousness as to be almost a pain to remember". It is adapted to a special kind of pollination. Pollen is vibrated out of horn-like anther appendages by bumble-bees beating their wings at a frequency below that in flight. Because of the sound, this is known as 'buzz' pollination. The adaptation seems successful as the pollen loads collected by bees visiting *Moneses* have been found to be very pure and not to contain pollen of other pyrolas which may grow in close proximity. However, if cross-pollination fails, *Moneses* has a fail-safe device. As the flower ages, the upward-pointing anthers are inverted by twisting of the anther filament which allows pollen to fall out of the pores in the horns, on to the stigma. However, the effectiveness of this mechanism is not known.

Moneses uniflora, Sutherland, 22 vi 1995.

The minute dust-like seeds, liberated through longitudinal slits in the spherical capsule, are similar to those of orchids. They are dispersed by wind and water and probably to some extent by animals. The seeds require a mycorrhizal fungus in order to germinate, but it is unlikely to be any of the same species associated with orchids.

Like many of our rare plants, *Moneses* has not become sufficiently familiar to people to be known by many local names. In Scandinavia, where it is common, the plant symbolizes the coniferous forest and names linked to most aspects of the plant's structure and growth habit have been coined. Perhaps most notable are 'porcelain flower' relating to the flower colour and texture of the petals; 'wood apple', one of the few common names that refers to the fruit capsule; and 'wurzel wanderer', meaning root wanderer, from the creeping thread-like roots which allow the plant to wander about the forest floor. The name 'St Olaf's candlestick' honours the patron saint of Norway, the 'candlestick' alluding to the projecting stigma.

Myosotis alpestris F. W. Schmidt

ALPINE FORGET-ME-NOT

In 1805 George Don published his *Herbarium Britannicum*, a collection of nine bound fascicles of pressed herbarium specimens. A range of rare Scottish alpine plants is represented and in the ninth fascicle is a specimen of *Myosotis alpestris* which constitutes the first record of the plant in Britain. Don described the plant as adorning the summit rocks of Ben Lawers as it still does today in lesser quantity. It was almost a further 50 years before the plant was discovered in Teesdale, its only other British locality. Here it was first noted on Mickle Fell by the two illustrious Yorkshire botanists, James Backhouse, father and son, in June 1852. Although *M. alpestris* has been recorded from a few hills in the vicinity of the original Scottish and English sites, the British distribution of this attractive mountain plant remains disjunct. It was collected from Caenlochan Glen by the Revs H. E. Fox and E. F. Linton in 1880 and recorded in the 1883 *Report of the Botanical Record Club*. However, this record was subsequently discounted. The Perthshire botanist Francis Buchanan W. White was informed by a horticultural friend that he had sown seed of cultivated plants from Ben Lawers, in the glen. The plant has also been listed for the Aonoch Mor area near Ben Nevis, but the record requires confirmation.

There are similarities but important differences between the habitats of *Myosotis alpestris* in Scotland and England. In both areas the plant is restricted to high altitudes where the winter temperature is generally below freezing and the summers cool, cloudy and wet. There is also a preference for southerly aspects where the plant is less likely to remain excessively wet. Like many alpine plants, *M. alpestris* is vulnerable to rotting if water is retained in the leaf rosette. In Britain the plant is only found on calcareous rocks; in Scotland it is confined to the soft mica-schist of the central Breadalbane mountains in Perthshire, and in England to the carboniferous limestone around the borders of Yorkshire and Westmorland.

The associated vegetation of *M. alpestris* differs in the two areas. In Perthshire the plant is a rarity in the rich dwarf-herb community usually dominated by sheep's fescue (*Festuca ovina*), alpine lady's-mantle (*Alchemilla alpina*) and moss campion (*Silene acaulis*). This high altitude community is characteristic of calcareous rocks on the lofty mountains of the central Highlands and it is the abundance of local and rare plants such as *Myosotis alpestris*, cyphel (*Minuartia sedoides*), rock whitlow grass

(*Draba norvegica*) and alpine cinquefoil (*Potentilla crantzii*) that characterizes this vegetation type. Equally characteristic of the carboniferous limestone of North Yorkshire and Westmorland is the community dominated by blue sesleria grass (*Sesleria albicans*) and *Festuca ovina* with limestone bedstraw (*Galium sterneri*), common wild thyme (*Thymus polytrichus* subsp. *britannicus*) and harebell (*Campanula rotundifolia*) constantly present. Here *Myosotis alpestris* can, in a few places, be very prominent in the vegetation. In both plant communities *M. alpestris* prefers areas of open vegetation which indicate that it avoids competition. The easily weathered mica-schist of the Breadalbane mountains constantly provides new niches for colonization whilst the turf on steep slopes at some of the English localities is kept open by soil creep. Scottish plants tend to have longer flowering stems and more upright leaves than those of Teesdale. The dwarfed nature of the English plants is thought to be an adaptive response to high grazing pressure on the limestone turf. As plants grown in cultivation from wild-collected seed retain this character, it appears to be a genetic response. Many of the Perthshire plants avoid grazing on inaccessible ledges.

Myosotis alpestris, Perthshire, 1 vii 1995.

The British localities of *Myosotis alpestris* constitute an interesting north-westerly outpost of the European distribution of the species. On the continent it is widespread in the southern mountain ranges which include the Pyrenees, Cantabrians, Sierra Nevada and the Alps. It also extends to the Apennines and east to Greece and Bulgaria. In central and southern Europe *M. alpestris* is part of a taxonomically complicated group of closely related species, but subsp. *asiatica* ranges eastwards to Kamchatka crosses the Bering Strait and continues across Alaska and reaches Colorado via the Rocky Mountains.

Myosotis alpestris is a very free-flowering, self-fertile species. These features are characteristic of a plant that relies on seed rather than vegetative propagation. The flowers produce nectar and several species of flies and butterflies have been recorded as visitors in Britain. The former are probably the most important pollinators in the high and exposed areas the plant inhabits. Flowering mainly occurs during June and July, and seeds (strictly the fruits) are shed during August and September. The black nutlets of *M. alpestris* are one of the most reliable characters by which to distinguish the plant from the closely related wood forget-me-not (*M. sylvatica*) which produces dark-brown nutlets. However, in Britain all known locations for *M. alpestris* are above the highest recorded altitude for *M. sylvatica*.

Nuphar pumila (Timm) DC.

LEAST WATER-LILY

Least water-lily was discovered in a lochan at the foot of Ben Cruachan in Argyll by William Borrer in 1809. In the British Isles *Nuphar pumila* is almost confined to the northern half of Scotland, with an outlying locality in both Kirkcudbrightshire and Shropshire. It has also been planted and is now classed as naturalized in Surrey. The species is much more restricted than the larger yellow water-lily (*Nuphar lutea*) and white water-lily (*Nymphaea alba*) both of which are found throughout Britain and Ireland, although the former is less common in the north of Scotland and the latter has been widely planted. Hybrid water-lily (*Nuphar* x *spenneriana*), a cross between *Nuphar pumila* and *N. lutea*, occurs in a number of sites in central and southern Scotland, with single populations in Northumberland and in Wales. All of these are remote from *N. pumila* and sometimes both parent species. The hybrid has a low seed fertility and is thus unlikely to have arisen in its isolated sites through seed dispersal. Most such isolated populations probably originated at a time when the ranges of the parent species overlapped more than they do today, with *N. pumila* occurring further south than at present.

Although the range of the two species overlaps in central Europe, *N. pumila* generally replaces *N. lutea* in the north and in the mountains, with its distribution extending northwards beyond the Arctic Circle in Scandinavia. *N. pumila* is classed as a circumboreal species, although there is a large gap in its distribution between eastern Siberia and central North America.

In Britain, *N. pumila* occurs from little over sea level up to 520m in Perthshire. It grows in sheltered inland waters with a low flow-rate and little turbulence. Habitats include ox-bow lakes, ditches, pools, marshes and bogs, with a muddy, silty or peaty substratum where the water is around neutral or only slightly acid. The water is often naturally very poor in nutrients, but in the oxygen-deficient upper layer of mud in which its rhizomes are rooted, nutrients, especially phosphates, are much more available.

Nymphaea alba, West Ross,
28 vii 1994.

Species of *Nuphar* have both submerged and floating leaves unlike *Nymphaea* which has only floating leaves. Connection between the leaves and their rhizomes ensures anchorage as well as a supply of nutrients and water. When water levels fall, leaves, which are exposed and surrounded by air are not damaged as long as they remain connected to rhizomes. The rhizomes of *N. pumila* are more slender than those of *N. lutea*, making it more vulnerable to disturbance caused by water fowl, boating and other recreations.

A system of intercellular air-spaces which gives buoyancy to the leaves also runs through the leaf stalks and supplies the roots with oxygen. The rootstock of water-lilies contains so much air that if pulled up out of the mud it will float and may be transported elsewhere. However, uprooted plants rarely get a chance to re-establish, except on exposed substrates at water margins, but here drying out or inundation often occurs before plants get a chance to take root. Thus, vegetative dispersal is relatively unimportant compared to seed dispersal for the spread of these water-lilies.

N. *pumila* occurs in communities of floating-leaved and submerged aquatic plants, sometimes as the dominant species in terms of water surface cover. It is frequently accompanied by pondweeds (*Potamogeton* species) in open-water and, where the shade is not too dense beneath lily pads, shoreweed (*Littorella uniflora*), bladderwort (*Utricularia*) and the naturalized Canadian waterweed (*Elodea canadensis*) may be found. Tall emergents can provide a sheltered microhabitat for N. *pumila*, reducing flow-rates as well as dissipating the energy of breaking waves. The marginal vegetation of shallower waters, especially peaty pools, often comprises stands of bulbous rush (*Juncus bulbosus*) and water horsetail (*Equisetum fluviatile*). In lime-rich sites with clear water N. *pumila* may be accompanied by a larger number of small aquatic plants including stoneworts (*Chara* species).

Nuphar pumila, Inverness-shire, 26 vii 1995.

In late summer, pools and lochans filled with least water-lily become polka-dotted yellow. Its flowers, similar to those of globeflower, *Trollius europaeus*, are only half the size of *Nuphar lutea*. Neither of the yellow water-lilies is as spectacular as the large white-flowered *Nymphaea alba* which decorates many a west Highland lochan. In both *Nuphar* species it is the yellow sepals which provide most of the colour, being much larger than the petals lying inside. As well as differences in flower size, the shape of the stigmatic disc in the centre of the flower is important for identification. It is lobed in N. *pumila* and entire in N. *lutea*. Further, the number of rays on the stigma (fewer in N. *pumila*) are diagnostic characters. As the flower opens, the stamens move outwards towards the petals, exposing pollen on their inner surfaces where it can be encountered by small insects such as flies and beetles. Their crawling within a flower often results in self-pollination, but insect flights to other flowers can effect cross-pollination. This relatively unspecialized pollination system does little to prevent

Flowers of Nuphar pumila (top) and N. lutea (bottom) showing stigmatic disc.

Nuphar x *spenneriana, Glasgow, 27 vii 1994.*

hybridization in sites where the two *Nuphar* species co-occur, although the somewhat later flowering of *N. pumila* reduces the chances of interspecific crossing.

The bottle-like fruit of the yellow water-lilies floats when detached from a plant. It soon breaks into segments (carpels) like an orange, each consisting of a dozen or more seeds surrounded by a mass of pulpy white tissue. Within two to three days, during which the segments may travel considerable distances by water currents and wind, the pulp rots or breaks off and the seeds sink to the bottom. The white fruit pulp is highly conspicuous and is eaten greedily by fish and water birds which may ingest the seeds unharmed. Experiments have shown that digestion of the seed-coat actually improves the germination capacity of the seeds. Birds are undoubtedly efficient agents of long-distance dispersal for *N. pumila* especially in mountainous districts, where the plant occurs in isolated lakes unconnected by drainage waters.

Orobanche alba Stephan ex Willd.

THYME BROOMRAPE

One may reasonably ask why our reddish thyme broomrape has such a seemingly inappropriate species name as *alba*. In 1807 James Edward Smith had in fact described and named the plant most fittingly *Orobanche rubra*. This was based on material sent by John Templeton who first discovered it on Cave Hill near Belfast in August 1805. However, in 1890, as a result of detailed study of the genus by the Bohemian botanist Professor Gunter Beck von Mannagetta, Smith's *Orobanche rubra* was considered to be merely a colour form of the widely distributed southern European species, *Orobanche alba*. On the continent the colour of *O. alba* is variable, with flowers ranging from red to yellow through to whitish, but the original description, published in Carl Ludwig Willdenow's *Species Plantarum* in 1800, describes the flowers as white only, hence the specific name *alba*. The description was obviously based on pale-flowered specimens without knowledge of the colour variation within the species, but because priority must be given to the earliest published name for a species *alba* takes precedence over *rubra*. Inappropriate as this is for the uniformly reddish plants in Britain, the name *Orobanche rubra* would be equally unsuitable for the whitish or yellow continental variants. As the strict rules of nomenclatural priority mainly apply only to botanical Latin names, Smith's common name 'red fragrant broomrape' may still be used.

In 1809, only four years after the discovery of *Orobanche alba* in Ireland, William Jackson Hooker and Dawson Turner, on an excursion to the Western Highlands and the Hebrides, found the plant on the island of Staffa. In the same year it was also recorded for the first time from a locality near Kirkcaldy in Fife where it still persists. A few years later, John Templeton, the original finder, encountered the plant again on the Giant's Causeway. It was also reported to Hooker from Salisbury Crags sometime before 1821 by the Edinburgh botanist R. K. Greville although, strangely, Greville did not record the plant from this locality in his *Flora Edinensis* of 1824. Most of these early records for *Orobanche alba* were from sites overlying basalt rocks. The apparent association of the *Orobanche* with this rock intrigued W. J. Hooker to the extent that in his account of the plant for *Flora Londinensis* in 1821 he states "If it were not a fear of subjecting myself to the imputation of making unnecessary changes in the nomenclature of the plants I describe, I should be very much disposed to alter the specific appellation of the one now under consideration, to *Orobanche basaltica*, thereby conveying to the student who knows the plant only in the herbarium, the information of its remarkable and peculiar choice of places of growth. As far as I have hitherto had an opportunity of investigating the point, my researches have confirmed me in the idea, that the *Orobanche rubra* confines itself wholly to basaltic rocks ...". Even though Hooker refrained from changing the Latin name, he offers Basaltic broomrape as a common name. Had the change been made, Hooker's name of *O. basaltica* would have suffered the same fate as Smith's *O. rubra* and it would have been proved not entirely appropriate as the plant has since been recorded from a wider range of rock types such as limestone in Yorkshire, western Ireland and Fife, Lewisian gneiss in Ross-shire and serpentinite on the Lizard Peninsula in Cornwall. Hooker also must not have been aware that the Kirkcaldy population grew on limestone rocks.

Orobanche alba, Cyprus, 18 iv 1994.

The distribution of *Orobanche alba* in Britain is distinctly western with its headquarters on the west coast of Scotland and the Hebrides. It is rare in the east which makes the very early discovery of the Kirkcaldy colony remarkable.

In England thyme broomrape has always been extremely localized and has become even more restricted through heavy collecting at some sites although it still survives on the Lizard Peninsula and on a few Yorkshire limestone scars.

Like all orobanches, *O. alba* is a root parasite, its main host in Britain being common wild thyme (*Thymus polytrichus* subsp. *britannicus*). Elsewhere in Europe, other hosts include a few other genera in the thyme family (Lamiaceae), such as *Salvia*, *Satureja* and *Origanum*, and also harebell (*Campanula rotundifolia*) and species of *Potentilla*, *Euphorbia* and *Heracleum*.

O. *alba* belongs to the group of plants that make up the Southern Continental Element of the British flora and is atypical in reaching the north of Scotland (its most northerly limit) and also reaching farthest north on the west side of the country. Most of the group do not extend to Scotland and reach their most northerly limit on the east side of Britain. However, thyme broomrape follows the pattern of the group on the continent with a generally southern distribution extending as far east as south-west Asia, with central Nepal its eastern limit and the north coast of Africa its southern.

The flowering period is from May to early September. Bumble-bees are attracted to the nectar-rich, fragrant flowers with their scent likened to honeysuckle or pinks. The pollination mechanism is similar in all *Orobanche* species whereby the four stamens are arranged in two pairs facing each other with the anthers of each facing pair held together like a pair of sugar tongs. The anthers, in the form of spoon-shaped receptacles, hold the pollen until forced apart by the bumble-bee in search of nectar. The pollen is then sprinkled over the head of the insect and is later deposited on the projecting stigma of the next flower it visits.

The abundant dust-like seed is wind dispersed but only germinates in the presence of a chemical stimulant released from the root of its host. The seeds must lie close to the host roots to receive the stimulant as it only penetrates the soil for about a centimetre. Moreover, once the radicle of the *Orobanche* seedling has grown to about 2mm, union with the host root is essential for further development.

Orobanche alba, West Ross.
(Photo: P Lusby).

Orobanche alba normally behaves as an annual, completing its life cycle within twelve months but sometimes may take longer to flower. It always dies after seed production with no means of vegetative propagation.

Oxytropis campestris (L.) DC.
YELLOW OXYTROPIS

Oxytropis halleri Bunge
PURPLE OXYTROPIS

Both species of *Oxytropis* that occur in Britain are rare and rank among the most attractive of our native alpines. The considerably rarer yellow-flowered *O. campestris* was discovered by George Don in 1812 in Glen Fee, one of the botanically-rich glens of the Clova mountains. This remains one of the plant's three known localities, the others being Loch Loch in Perthshire, and the Mull of Kintyre. At this last site the plant was originally mistaken for *O. halleri* because of the purple tinge to the pale yellow flowers.

O. halleri was first found in 1761 near Loch Leven by the Rev. Dr John Walker, Professor of Natural History at the University of Edinburgh. This date is early for first records of Scottish plants; most of the country, especially the Highlands, was still largely unexplored by botanists. One of the pioneers in the latter half of the eighteenth century was James Robertson, a student of John Hope, whose tour of 1767 took him (fortuitously) to most of *O. halleri's* other sites, on the coasts of Sutherland, East Ross and Argyll. Apparently unaware of Dr Walker's find, Robertson published his own observations in the *Scot's Magazine* in 1768 of "a new species". Had Robertson taken the ferry to North Queensferry instead of Kinghorn on his journey north he might also have recorded *O. halleri* from Fife. The plant was collected in numbers from around this latter locality in the nineteenth century by, among others, Professor John Balfour and his botanical students from Edinburgh. Any surviving plants were finally lost when the site was destroyed by construction of the railway cutting for the Forth Bridge in the 1880s. *O. halleri* has also disappeared from other sites in Fife, Angus and Caithness. It can still be found on two mountains in Perthshire and one in Argyll, and also on the coast of the Mull of Galloway, Cromarty and, most abundantly, in Sutherland.

Oxytropis campestris, Angus, 10 vii 1995.

Oxytropis campestris, Angus, 17 vi 1982. (Photo: P Lusby).

O. campestris descends to near sea level in Kintyre, where it grows on the face of an exposed limestone sea-cliff together with other alpines such as mountain avens (*Dryas octopetala*) and hoary whitlowgrass (*Draba incana*). However, it is more of a montane plant than *O. halleri* in Britain. The frequency of the latter on low coastal limestone hills and calcareous sand dunes, not just on the extreme north coast, but in Ross-shire and formerly in Fife and Angus, is typical of a montane-coastal distribution. Other species which grow on mountains inland and by the sea include common scurvygrass (*Cochlearia officinalis*), thrift (*Armeria maritima*) and sea campion (*Silene uniflora*). These plants were noted by Robertson who was among the first to observe and comment on this unusual type of distribution. Like *O. halleri*, they are tolerant of high exposure to wind and sun and thrive on a free-draining basic soil. Whereas their leaves are fleshy, those of *O. halleri* are covered with silky hairs, both features primarily an adaptation to drought. *O. halleri* is additionally a distinct calcicole, growing on basic igneous rocks in Argyll, mica-schist in Perthshire, (formerly) calciferous sandstone in Fife, and calcareous shell-sand in Sutherland. In the last-named locality the

Oxytropis halleri, Sutherland,
16 vi 1995.

plant grows in herb-rich *Dryas octopetala* heath with several arctic-alpines almost at sea level – the purple and yellow saxifrages (*Saxifraga oppositifolia* and *S. aizoides* respectively), hoary whitlowgrass (*Draba incana*) and moss campion (*Silene acaulis*).

O. campestris also behaves as a calcicole. In Glen Fee it grows on a distinct strip of whitish micaceous rock, where among its neighbours are alpine lady's mantle (*Alchemilla alpina*) and roseroot (*Sedum rosea*). Both *Oxytropis* species are regarded as climatic relics in Britain and were probably more abundant when open and base-rich habitats had not yet been colonized by forest or acidified through leaching and the development of bog and heath. Seeds of *O. campestris* have been found in the stomachs of ice-preserved mammoths which roamed widely during the Pleistocene in southern Britain, Europe and northern Asia.

O. halleri belongs to the Alpine Element of the British flora which occurs in the mountain ranges of central Europe (the Alps, Pyrenees, Carpathians and northern Balkans), but not in the Arctic. The relative rarity of the two species of *Oxytropis* is reversed outside Britain. In addition to central Europe, *O. campestris* is found in south-east Sweden, including the limestone islands of Oland and Götland, Finland, northern Russia, the Alps, across northern Asia and in north-east and western North America.

Apart from differing in flower colour, *O. campestris* is generally larger than *O. halleri*, but large plants of both can be found in especially favourable spots. The seed-pods of *O. halleri* are slightly longer than those of *O. campestris* and clothed with brown, close-appressed hairs, while those of *O. campestris* also have short spreading hairs. *O. halleri* could be confused with purple milk-vetch (*Astragalus danicus*) or alpine milk-vetch (*A. alpina*), but the sharp-pointed keel petal of *Oxytropis* is a reliable feature. The name *Oxytropis* is derived from the Greek, *oxus* (sharp) and *tropis* (keel).

While some species of *Oxytropis* are palatable to grazing animals, others are regarded as the most deadly to livestock of all poisonous plants, whence the names locoweed, crazyweed and poison vetches. *O. campestris* is known as yellow locoweed in North America. Symptoms of locoweed poisoning include dragging of the feet, chomping of the jaws and lack of muscular control; afflicted animals literally go "loco" (Spanish for crazy). Locoweeds contain a toxic alkaloid called swainsonine which is present in very low concentrations in the leaves. However some species also accumulate selenium from the soil, an element especially high in sedimentary rocks and limestones. Scottish plants, at least of *O. halleri*, are presumably non-toxic as they are heavily grazed by sheep or goats at some sites. James Robertson, hoping to collect seed from a population of *O. halleri* he discovered near Tongue was "very near being disappointed, as the cattle scarcely had left any of it"; and at nearby Farr Bay the plant was "everywhere eaten down by the cattle ... even in preference to the white clover [*Trifolium repens*] among which it grows".

Robertson's chief purpose was to record the plants he found, but his journal also contains notes on the agricultural potential of many native species. Of *O. halleri* he wrote, "it promises to be an useful early and hardy perennial grass ... [which] when made into hay has an agreeable, astringent herbaceous taste", but he pointed out that the plant would not withstand repeated cutting because its leaves grow directly from the root. Robertson considered the plant could perform a dual function on windswept sandy coasts, of cattle fodder and a sand-stabiliser. Neither *Oxytropis* is thus employed in Britain. However, the species' silky leaves and attractive flowers have won them a well-merited place in alpine and rock gardens.

Both species are free-flowering when ungrazed and produce nectar which is collected by bees. These are the main pollinators, but they sometimes act as nectar-robbers by biting a hole through the base of the corolla without contacting either stigma or anthers. Flowering time is similar in both species, between May and late July. For *O. halleri* it varies with the shortness of the growing season, being delayed by up to two months on the north coast, compared with its nearest site in Cromarty.

Oxytropis halleri on sand dune, Sutherland, 15 vi 1995.

Phyllodoce caerulea (L.) Bab.
BLUE HEATH

James Brown is usually credited with the discovery of *Phyllodoce caerulea*, on "a dry moor in the district of Moray near Aviemore" in 1815. However, James Robertson recorded a plant which he thought was new to Britain and "not mentioned by Linnaeus or any other botanist" while crossing "the mountains that ly at the head of the water of Findhorn" on 26 June 1771. Robertson initialled his discovery "A...C..." in his journal probably as a way of keeping it secret, as he did "S...C..." for *Saxifraga caespitosa*, until he and James Hope had a chance to publish them. The fact that *Phyllodoce caerulea* (=*Andromeda caerulea*) had already been described by Linnaeus is rather puzzling. However, Robertson's route from Pitmain, by Kingussie, to Strath Dearn, probably did take him not far north of the only area where the plant is now known.

The seven or so known sites lie no more than 15km apart, on the Sow of Atholl and the Ben Alder range, west of Loch Ericht. Several of these sites have been found within

Phyllodoce caerulea,
Inverness-shire, 3 vii 1995.

the last 40 years, and as long as botanists roam the high hills it is possible that its distribution will be extended. Herbarium collections in the 1830s from Strathspey would appear to confirm Brown's 1815 record, though the exact whereabouts of the plant's locality or localities here are not known. There is no specimen substantiating George Don's record from the Western Isles of Shiant. A search by Charles Babington in 1841 failed to find the plant, convincing him of its absence or an error of the locality.

Phyllodoce caerulea has a distribution best described as circumpolar, but with large gaps in eastern and central Asia and western North America. In central Europe it is present only in the Pyrenees, and in Europe as a whole it is only really common in Scandinavia. It is both arctic and alpine, growing on snowy arctic coasts, especially in Greenland and Norway, including the Lofoten Islands, and in the Kjølen mountains it reaches 1850m on Jotunheimen.

Throughout its range, *Phyllodoce caerulea* is chionophilous, literally 'snow-loving'. Its restriction to sheltered hollows and predominantly north-facing slopes where snow melts as late as June or July is one reason why the plant has eluded botanists for so long in some sites, and why visitors to known sites are not always ensured a sighting of the plant. It is also a distinct calcifuge, unlike many of our rare alpines which are confined to basic rocks. The vegetation thus provides little hint of its whereabouts or recompense for fruitless searches by way of a species-rich calcicolous flora.

The colonization characteristics of *Phyllodoce caerulea* on a glacier foreland in Jotunheimen agree with observations of the plant's ecological behaviour in Scotland. It prefers well-drained sites, low in nutrients, and is absent from ground where the pH is greater than five. Plants of *P. caerulea* in the colonization zone nearest the glacier snout generally look unhealthy, probably because the most recently exposed till is still quite basic. Ageing of the ground, which includes a lowering of the pH, is necessary before the plant forms regenerating populations.

Phyllodoce caerulea is rather a poor competitor, relying mostly on seedling establishment to increase its area of occupation. An average of 18 seeds per seed capsule has been found in the Scottish populations, with a maximum recorded number of 25, a rather meagre total considering their minute size. The seeds are doubtless spread by wind and probably water, but ptarmigan, which eat off the seed capsules, are a more promising means of direct dispersal to suitably sheltered snowy hollows at high altitude. Seedlings appear to be rare in Scotland and it is likely that conditions necessary for establishment are infequently met, not least due to untimely frosts and summer drought. Layering is favoured in better drained sites, but the formation of new plants by this method of propagation is also infrequent. However, perhaps by way of compensation, established plants are very long lived; ages of at least 60 to 70 years have been determined by ring counts of woody stems in Scotland and Norway.

The plant's preference for snow-bound sites agrees with the fact that the shoots are not totally frost-hardy. Exposed shoots can be killed by early frosts or damaged by blown ice crystals. Beneath a covering of snow, shoots are not only protected from frost, but can initiate new growth in the relatively warm, moist microclimate. The life-span of leaves can be shortened to two to three years in exposed sites, compared with four years in sheltered hollows.

Despite its name, *Phyllodoce caerulea* is not a true blue heath; indeed one is not known. Its flowers have more of a pinkish hue due to being dotted with red glandular hairs. The bluest part of the plant is its foliage which is not unlike that of a small densely-leaved *Hebe*. Vegetatively *Phyllodoce* resembles mountain crowberry (*Empetrum nigrum* subsp. *hermaphroditum*), and could easily be mistaken for this plant which is one of its commonest associates in snowy hollows. The former specific name for *P. caerulea*, *taxifolius*, was given because the shoots resemble those of a miniature yew tree (*Taxus* = yew).

Last century the plant was popularly known as Menzies heath or Scottish Menziesia, a name which lives on as the plant is the official badge of Clan Menzies. The American heath genus *Menziesia* honours Perthshire-born Archibald Menzies, best known for his introduction of plants from the American continent to Britain, including monkey puzzle (*Araucaria araucana*). Menzies' name has been attached to a large number of plants, many of them grown in British gardens. *Phyllodoce caerulea*

Phyllodoce caerulea, Perthshire, 28 vi 1995.

can be grown successfully in gardens even in the south of England where there is rarely any snow, but equally little chance of frost.

Since being named *Andromeda caerulea* by Linnaeus, the plant has been included in three other heath genera, *Erica, Menziesia* and *Bryanthus*, before Charles Babington established it as a member of *Phyllodoce*.

Phyllodoce was a nymph mentioned by Virgil in the Georgics, whose "glossy locks o'er snowy shoulder shed" while she spun "wool stained with hyaline [glassy blue] dye". Thus it could be said that both *Phyllodoce* and *caerulea* describe the 'blue' flowers, while the former also suggests the plant's preferred habitat.

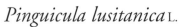

Pinguicula lusitanica L.

PALE BUTTERWORT

In 1662, John Ray discovered in Cornwall "a small sort of *Pinguicula* which seems to differ specially from the common". He described it in his *Catalogus Plantarum Angliae* of 1670 as a species new to science, *"Pinguicula flore minore carneo"* [butterwort with a small, flesh-coloured flower]. Soon afterwards the same plant was found in Devon and later "on the bogs of Dorset" by Richard Pulteney in 1765. John Hope and John Lightfoot extended its range to Scotland, by finds in Ayrshire and Arran, and the Isle of Skye, respectively, and recorded it as *Pinguicula villosa* (hairy butterwort). It was not until 1794 that the identity of the British plants was established when J. E. Smith and J. Sowerby compared specimens with Portuguese material of *Pinguicula lusitanica*.

The alternative common name for pale butterwort, western butterwort, accurately describes the British distribution of the species. In Scotland it mainly occurs in west Sutherland, West Ross, Inverness-shire, Argyll and the Hebrides, but extends south to the coast of Galloway. It also occurs on the Isle of Man, the Welsh coast of Pembrokeshire, and around the south-west coast of England from Somerset to Cornwall and Hampshire. It is scattered throughout Ireland, but is most frequent in the west.

P. lusitanica is a representative of the Lusitanian Element in the British flora, which is composed of plants having their main area of distribution in Spain, Portugal and round the western Mediterranean. The name derives from 'Lusitania', the ancient name for the western part of the Iberian Peninsula. The range of the Lusitanian flora extends northwards along the Atlantic seaboard as far as Ireland, a route along which most species probably migrated after the last glaciation when the climate was warmer and the coastline was continuous with southern Europe. However, *P. lusitanica* is less typical of the Lusitanian Element than its name suggests because of its extension so far north, reaching its limit in Orkney. A more typical distribution is that of its cousin, large-flowered butterwort (*P. grandflora*), which reaches only western Ireland.

One unifying feature of the Lusitanian plants, which explains their southern and western tendancy in the British Isles, is their intolerance of frost. *P. lusitanica* requires a frost-free and relatively long growing season to flower and set seed. The plant's other major requirement, high humidity, is largely provided by its local habitat. Its preference for south- or south-west-facing aspects is maintained or even sharpened in the south of its distribution, despite the dryness of the regional climate. This can be accounted for by the fact that only on sunnier slopes is there sufficient vaporization of water to maintain locally high humidity.

The carnivorous *P. lusitanica* typically grows on wet, open peat or mineral soil which is constantly irrigated by slight to moderately base-rich ground water, or on various substrates of impeded drainage, but where inundation is nevertheless uncommon, except in winter. In wet heaths or mires, *P. lusitanica* occurs in more or less isolated flushes with other weakly-competitive plants. Its associates include tawny sedge (*Carex hostiana*), flea sedge (*C. pulicaris*), bulbous rush (*Juncus bulbosus*), bog asphodel (*Narthecium ossifragum*), yellow saxifrage (*Saxifraga aizoides*), and species of sundew (*Drosera* spp.). It is rarely found within taller vegetation, and then only in gaps, such as between clumps of bracken (*Pteridium aquilinum*) and bushes of bog myrtle (*Myrica gale*). The size of single populations is limited by the size of such gaps or flushes, but is mostly between ten and 200 plants.

Pinguicula lusitanica, West Ross, 6 vii 1995.

Although the leaf rosettes are small and very pale, sometimes almost transparent, they stand out against the dark soil like little stars, and on closer inspection their net of thin red veins is conspicuous. In particularly sun-exposed habitats, the whole leaves may acquire a pinkish hue. Pale butterwort differs from most other *Pinguicula* species of temperate climates in producing only one type of leaf, and overwintering as a rosette instead of forming a winter resting bud. The latter, known as a hibernaculum, is rootless and can be dislodged by frost action, water or animals which may thus assist in dispersal. In addition, most other butterworts produce even smaller winter buds, or gemmae, which afford a means of propagation often superior to seed recruitment in maintaining populations. The lack of vegetative propagation in *P. lusitanica* means that the species is entirely dependent on establishment from seed, a feature which corresponds with its climatic and aspect restrictions in Britain. Late spring or early autumn frosts are particularly damaging to the plant's reproductive capacity.

The wide, pale lilac-pink flower with red-streaked yellowish throat resembles a small *Gloxinia* and is pollinated mostly by small flies. These are among the victims trapped by the leaves, in addition to winged aphids, spring-tails and a variety of tiny creeping insects. Seldom is anything the size of a house-fly retained, but the small size of the prey is compensated by their large numbers. Not only insects are trapped; pollen, seeds and leaf fragments can augment a conventional insect diet, pollen in particular being a good source of nutrients.

The carnivorous nature of the plant was first proven by Charles Darwin who carried out experiments showing the leaves to digest and absorb prey. The name butterwort comes from the greasy feel of the leaves, as does the Gaelic name, Measgan 'a little butter dish'. *Pinguicula* is derived from the Latin *pinguis*, meaning fat or greasy. Tiny stalked glands covering the surface of the leaf produce mucilage, making the leaves greasy. The leaves have long been known to curdle milk, reflected in the Manx name, Lus y steep, and Lus a bhainne, milkwort. This property is due to digestive enzymes contained in the leaves' secretions. Five different enzymes are produced, mostly by glands which are excited into activity by contact. The combined action of these enzymes and acid contained in the mucilage rapidly dissolves and digests the soft parts of an insect, whose nutritive juices are absorbed by the leaf.

This supplementary form of nutrition benefits the plant in nutrient-poor habitats, where the supply of nitrogen in particular is very low. However, the leaves are costly to produce and maintain, so *P. lusitanica*, like other carnivorous plants, is constrained to growing in open habitats where there is ample light and warmth for photosynthesis.

P. lusitanica has lost many sites, especially in the southern lowlands, due to drainage and various forms of land use including peat-extraction, moor burning and, more recently, afforestation. Another threat is eutrophication from agricultural sources, which may stimulate the growth of tall, lush vegetation. The plant is also susceptible to trampling and disturbance, more so than common butterwort (*P. vulgaris*) which completes its flowering earlier and is not entirely dependent on seed-set for reproduction.

Pinus sylvestris L.

SCOTS PINE

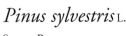

Pinus sylvestris is the best known of Scottish trees. Its beauty, viewed from every angle in the changing light of the day and throughout the seasons, has inspired many a writer, artist and photographer.

Scots pine has been present in Britain in all the interglacial phases of the Quaternary period (the last 1.6 million years) and its distribution has expanded and contracted in response to the changing climate. The tree gained prominence during cooler periods in the early and late stages of interglacials, but gave way to broadleaved trees in intervening warmer temperate phases.

Opinions differ as to whether our native Scots pine reinvaded Britain from the continent after the last glaciation or survived and spread from isolated pockets called refugia. Genetic research and studies of fossil pollen have led some researchers to conclude that native Scottish populations of *P. sylvestris* originated from more than one glacial refugium in Scotland (and possibly Ireland). Pine forests as such would not have survived but groups of trees may have. Some former English populations may have had a continental origin.

Pinus sylvestris spread throughout the British Isles as conditions ameliorated after the Late-glacial period and reached its maximum abundance during the cool, dry Boreal period (6000–5000BC). With increased warmth and moisture during the Atlantic period (5000–3000BC) it declined in favour of broadleaves and by the Sub-Boreal (3000–500BC) had become scarce in comparison with oak and had almost disappeared from Ireland. Man then began to influence the vegetation. By about 500BC few Scots pine survived in England, Ireland and Wales.

From Roman times to the the end of the 16th century demand for timber was concentrated in the lowlands. It was not until about 1600, when the demand for timber became acute, that attention was drawn to the Highland forests by early travellers. It was only then that the first botanical record of the pine was made by the herbalist John Parkinson in his *Theatrum Botannicum* of 1640. The following two centuries saw the main destruction of the Caledonian pine forests with devastation in some woods by the middle of the 18th century. Timber was primarily used for smelting as well as by the navy (especially for masts), but it was also important in construction. Other uses included barrels for the herring fleet, boxes and oars. In north-west Scotland tripods of split pine (stack centres) were used to dry hay and corn and pine strips were used for household lighting in a similar fashion to rush lights. The durable Rothiemurchus pines were bored to make water pipes which were exported to London.

Pinus sylvestris, Aberdeenshire, 11 vii 1995.

Difficulties of access and extraction slowed the pace of forest clearance, although once the techniques of floating logs down the Highland rivers were perfected, large areas were harvested. The steady increase in sheep and deer stocking prevented natural regeneration. Major exploitation of many Highland pinewoods had ceased by the end of the 19th century but during World War I, and to a lesser extent in World War II, significant areas were felled in several of the larger remnants of the pine forest.

The fragmented remains of the Caledonian forest extend north from Glen Falloch in south-west Perthshire to Glen Einig in West Ross, and east from Shieldaig on the coast of West Ross to Glen Tanar on Deeside. Isolated trees reach an altitude of 840m on Ben Macdui but the upper limit of pinewood reaches 615m on Creag Fhiaclach in Rothiemurchus where the stunted, twisted trees probably form the only natural altitudinal treeline in Britain.

Native pinewoods are not rich in associated flowering plants. The ground vegetation is usually dominated by ericaceous dwarf-shrubs such as heather (*Calluna vulgaris*) and a variable mixture of blaeberry (*Vaccinium myrtillus*) and bell heather (*Erica cinerea*) with wavy hair-grass (*Deschampsia flexuosa*) and several common mosses including *Hylocomium splendens*, *Dicranum scoparium*, *Pleurozium schreberi*, *Rhytidiadelphus triquetrus* and *Hypnum jutlandicum*. The wetter western pinewoods are characterized by species of *Sphagnum* and oceanic bryophytes. In the drier eastern pinewoods various bulky mosses are abundant. Where pinewoods extend onto damp, peaty soils cross-leaved heath (*Erica tetralix*) and purple moor-grass (*Molinia caerulea*) become prominent and on fertile soils hairy wood-rush (*Luzula pilosa*) and wood sorrel (*Oxalis acetosella*) increase. The range of plants associated with pine in Scotland is poorer than in Scandinavia. Populations of plants will very likely have been lost, and the variety of habitats reduced by the widespread clearance. Plants such as interrupted clubmoss (*Lycopodium annotinum*), bog bilberry (*Vaccinium uliginosum*) and twinflower (*Linnaea borealis*) commonly occur in Scandinavian pinewoods but are absent or scarce in Scottish pinewoods.

Scotland is an oceanic north-western outpost for *Pinus sylvestris*. It is one of the widest ranging conifers extending some 14,000km, from north-east Spain to the Pacific coast towards Kamchatka. It reaches north of 70°N in Scandinavia and south to 37°N in the Sierra Nevada. Low rainfall is limiting at both latitudinal margins while low temperature prevents seed ripening at its northern limit. Native Scots pine grows mainly on freely drained acidic soils of glacial origin but in parts of Europe and Asia it occurs on a variety of other soils, from calcareous earths over limestone to acid peaty soils with thick humus layers, and even deep spongy bogs. More than a hundred fungi are recorded as forming mycorrhizal associations with *Pinus sylvestris*. This is no doubt an important factor which enables Scots pine to grow on a wide range of soils but it will not tolerate heavy shade.

P. sylvestris has been present in Europe since the beginning of the Tertiary period (65–1.6 million years ago). It is a highly variable species, perhaps a reflection of its long history and wide range. Over 100 variants have been described based on characters such as shape of the crown, branching pattern and the size, colour and shape of the needles, seeds and cones. Bark texture and colour and quality of wood are also important. Indigenous Scottish trees have been distinguished as the endemic subspecies or variety *scotica* chiefly by the shape of the crown, which remains pyramidal until late in life when it becomes rounded. This feature has, however, been observed elsewhere in its range.

The remaining stands of native pine have assumed great importance for nature conservation and great emphasis has been placed on maintaining their genetic integrity. To this end trees known to be of foreign origin are removed from natural populations and management to extend existing native stands is generally by natural regeneration or planting from seed of local origin. However, it has been shown that there is still considerable genetic diversity within and between native populations of *Pinus sylvestris*. It is not possible to know how this would compare with the former pinewoods since the forests were selectively depleted, with the most vigorous, straight-stemmed specimens being favoured. Since the 17th century Scots pine of mixed origin has been planted in the vicinity of native stands and, as *P. sylvestris* is an outcrossing species with pollen and seed adapted for long distance wind dispersal, many Scottish populations have probably hybridized to some degree with imported trees. Reduced seed development has been observed in small, isolated stands, with high levels of self-fertilization, as at Glen Falloch. Management to increase natural regeneration by reducing deer grazing is being increased at this site, but a greater genetic diversity may be required to increase seed production.

Polygonatum verticillatum (L.) All.
WHORLED SOLOMON'S-SEAL

The indefatigable George Don was the first to find whorled Solomon's-seal in Britain, in a wooded den east of Dunkeld, Perthshire in 1789 or 1790. This site is one of nine known to survive in Scotland, all within the River Tay catchment. At least two documented Scottish sites have been lost since last century, and the plant's only English locality, on the North Tyne, was last recorded in 1866.

In Scotland *P. verticillatum* is at the western edge of its Eurasian distribution which stretches east as far as the Himalayas. From about 70°N in Norway its range extends southwards through western Scandinavia, Denmark and most of Central Europe to

Polygonatum verticillatum, Perthshire, 26 vi 1995.

the Pyrenees, Apennines, Carpathians and Caucasus. It is increasingly confined to the mountains in southern Europe in accordance with the plant's primary requirement, high humidity, and its intolerance of extreme summer drought. Over much of Europe its distribution is to some extent relictual, resulting from the widespread loss of lowland deciduous forest.

In Scotland *P. verticillatum* is virtually confined to ancient woodlands which have survived in a near-natural state, protected from intensive exploitation by domestic grazing and woodland management mainly through their inaccessibility. Eight of the nine sites are wooded gorges or stream valleys, the other being a narrow strip of riparian woodland. The plant shows a preference for base-rich soils but will grow in areas of acid geology where flushing or periodic inundation provides an adequate supply of nutrients. In western Scandinavia, where it occurs widely up to the forest limit, it is rarely found in spruce or pine forest except on basic rocks where the soil resembles a brown earth typical of lowland deciduous woods. In south-west Sweden the plant is associated with wooded meadows, now mostly abandoned, where it survived the traditional spring burning of litter by virtue of its underground rhizomes, and also benefited from the reduced competition of other tall herbs and grasses through late-summer mowing. The plant communities of these wooded meadows are adapted to low-intensity management maintained over many centuries, which in certain respects parallels the long-continuity of the field layer of ancient woodlands.

In its Scottish sites *P. verticillatum* grows with a number of other herbs of base-rich woodlands such as herb paris (*Paris quadrifolia*), dog's mercury (*Mercurialis perennis*), marsh hawk's-beard (*Crepis paludosa*) and sanicle (*Sanicula europaea*). The plant's most characteristic habitat in Scandinavia, sub-alpine woodland, has all but disappeared from Scotland. Here it grows in luxuriant tall-herb vegetation with alpine sow-thistle (*Cicerbita alpina*), meadow crane's-bill (*Geranium sylvaticum*), meadow-sweet (*Filipendula ulmaria*), raspberry (*Rubus idaeus*) and northern monkshood (*Lycoctonum septentrionale*).

Polygonatum verticillatum, Perthshire, 13 vi 1995.

Fruit and autumnal foliage of Polygonatum verticillatum, Perthshire, 14 x 1994.

P. verticillatum fruits rather poorly in Scotland, one of the reasons being the lack of opportunity for cross-pollination. The small size and isolation of the surviving populations means the majority of flowers are unpollinated or pollinated by pollen from a neighbouring shoot of the same clone which does not always result in fruit formation. Its populations are maintained through vegetative propagation and the rhizomes, being very long-lived, allow the plant to survive in conditions which do not favour sexual reproduction. Although very shade-tolerant, *P. verticillatum* scarcely flowers under a closed canopy and light seems to be critical in promoting flowering.

When fruits are produced they often remain on the shoots until the leaves senesce in autumn when the scarlet berries, contrasting with the yellowing foliage, may be more conspicuous to birds. Its autumn colour was certainly a feature that attracted gardeners to the plant. Whorled Solomon's-seal has been grown in British gardens, originally as a medicinal herb, since the 16th century and possibly earlier. It was most popular in the late 1800s as a hardy perennial for the herbaceous border or woodland garden, and was also forced in early spring for conservatory decoration. The experience of gardeners has provided useful information about the plant, the biology of which is poorly known. Rhizomes forced too frequently failed to flower because their supplies of stored materials became exhausted, and division of rhizomes into too small sections resulted in the buds on the rhizomes aborting. Established populations of the plant are intolerant of soil disturbance which fragments the rhizomes, and it has been found that there is a minimum rhizome size below which flowering shoots are not produced. Nutient-rich soil encourages the growth of larger rhizomes, and in combination with open to semi-shade conditions produces the most vigorous and free-flowering populations.

Primula scotica Hook.
SCOTTISH PRIMROSE

Any book on Scotland's plants would be incomplete without this little gem, found only in the north of Scotland and nowhere else in the world. *Primula scotica* is one of the very few plants unique to the British Isles and one of even fewer unique to Scotland.

P. scotica was at first mistaken for *Primula farinosa*, bird's-eye primrose, when James Robertson found it near Tongue in Sutherland in 1767. A number of botanists, amongst them John Hope and John Lightfoot, also believed it was the same species or a northern or dwarf variety of it. Not until 1819 was *Primula scotica* described as a distinct species, by W. J. Hooker in *Flora Londonensis*. According to Hooker it had for some time been grown in gardens under that name and was then in cultivation at the Royal Botanic Garden Edinburgh. The name *Primula scotica* may have been originally given by the Ayrshire nurseryman, James Smith, who offered the species for sale in 1827 when he wrote, "it was I who named it, and, I believe, first detected it as a new British plant". Smith was first shown the plant in about 1808 by John Dunlop who believed he had found *Primula farinosa*, but Smith remarked, "if not a new species, it was a singular variety".

Primula scotica, Sutherland, 18 vii 1995.

P. scotica is distinguishable from *P. farinosa* by its smaller size, shorter and broader leaves with smooth rather than serrated margins, shorter and more inflated calyx, and its deep purple, not pink, flowers. *P. farinosa* is now extinct in Scotland, though it formerly occured in Peeblesshire and Lothian. The origin of *P. scotica* is still uncertain but one theory is that it arose from a cross between *P. farinosa* and its closest relative *P. scandinavica*. Fossil remains of *P. scotica* and a species similar to *P. scandinavica* have been found in a Cambridgeshire peat bed. Therefore the species may have arisen farther south of its present-day distribution and is likely a post-glacial relict in Scotland.

Primula scotica – form with
very short flowering stem,
Sutherland, 18 vii 1995.

P. scotica is restricted to the north coast of Scotland between Cape Wrath in Sutherland and Dunbeath in Caithness, and Orkney where it occurs on most of the larger islands. All its sites are within one and a half kilometres of the sea. Although formerly found farther inland, it has been lost from these sites due to increased agricultural intensity. The plant has very specialized ecological requirements and is restricted to four main types of habitat: calcareous sand dunes or machair, coastal limestones such as at Durness, maritime heath and grassland. Although under maritime influence, indicated by species such as sea plantain (*Plantago maritima*), spring squill (*Scilla verna*) and thrift (*Armeria maritima*), the vegetation zone to which *Primula scotica* is confined lies inland of the zone of highest salt exposure, but has a seaward limit coinciding with that of the dwarf shrub and herb communities. The failure of *P. scotica* to spread farther inland is due to its inability to compete with much taller vegetation. All its sites are grazed by rabbits and/or sheep which help to keep the vegetation short and open. When the rabbit population was drastically reduced by myxomatosis *P. scotica* disappeared or declined in several sites.

Species found with *P. scotica* in grazed coastal grasslands on base-rich substrates are many and varied, sometimes up to 50 species in an area of four square metres. They include small herbs such as field gentian (*Gentianella campestris*), frog orchid (*Coeloglossum viride*), bird's-foot trefoil (*Lotus corniculatus*), grass-of-Parnassus (*Parnassia palustris*) and heath milkwort (*Polygala serpyllifolia*). In a few places *P. scotica* occurs in a basic mire dominated by short sedges and species such as black bog-rush (*Schoenus nigricans*), lousewort (*Pedicularis sylvatica*), cross-leaved heath (*Erica tetralix*) and a long list of liverworts and mosses. *P. scotica* requires soils that are reasonably well drained but which never dry out completely.

Botanists have followed the fortunes of *P. scotica* populations over several years and found the number of plants to fluctuate considerably. In years with low spring and summer temperatures and summer gales, flowering and seed-set are reduced, with a corresponding reduction in the number of seedlings appearing the following year. Plant survival is favoured by mild winters, but because plants do not flower until they are two or more years old it can take several years for a depleted population to recover former levels of reproduction. Extremely low levels of genetic variation have been found within and between populations of *P. scotica*, raising concern that the species will not be able to adapt to climatic changes. Contrary to the general trend of climatic warming, at latitudes above 50°N winters have become colder since the 1950s and strong winds more frequent, conditions which do not favour the survival of the plant.

P. scotica is also sensitive to changes in grazing pressure. If grazing becomes either too severe or too relaxed populations will decline and may become extinct. On its reserve on Hoy in Orkney the Scottish Wildlife Trust has been experimenting with the use of Shetland sheep to try and encourage numbers of *P. scotica*. The project has been successful so far with one population more than doubling in size. Management of the surrounding land, not just that supporting *P. scotica* may also affect the survival of the plant. Sheep pastured on fertilized grassland which are allowed to wander in the area of *P. scotica* can enrich the vegetation resulting in a taller sward and increased competition. The future of Scottish primrose in agricultural areas is very much dependent on its sites remaining in suitable condition through careful management.

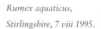

Rumex aquaticus L.
SCOTTISH DOCK

The discovery of true *Rumex aquaticus* in Scotland in 1935 finally laid to rest a fascinating problem that had confused British botanists for over 200 years. Difficulties commenced when Linnaeus described *Rumex aquaticus* in his *Species Plantarum* of 1753 but omitted to describe a closely related plant which is now known as *Rumex hydrolapathum* (water dock), although he was probably familiar with it. Linnaeus' description of *Rumex aquaticus* is extremely short but specifically draws attention to the fruit characters that distinguish it from the water dock *"Valvulis integerrimis nudis ..."* [fruit valves entire and bare, that is, without tubercles]. The main differences between the Scottish dock (*R. aquaticus*) and the water dock (*R. hydrolapathum*) are that the former has large triangular basal leaves and fruits with no protuberances (known as tubercles) on them, whilst the latter has broadly lance-shaped basal leaves and fruits with tubercles. Linnaeus seemed to have included both the Scottish dock and the water dock under the one name of *Rumex aquaticus*.

Rumex aquaticus,
Stirlingshire, 7 viii 1995.

William Hudson of Kendal was first to realize that Linnaeus' description was not applicable to the water dock and so he provided the name of *Rumex hydrolapathum* for it in the 1778 edition of his *Flora Anglica*. Unfortunately, this was too late, for in the previous year John Lightfoot in his *Flora Scotica* had already misapplied the Linnean description and name of *Rumex aquaticus* to the water dock. This caused confusion until James Edward Smith pointed out the mistake in his *Flora Anglica* of 1824. Smith admitted that even he had confused *Rumex aquaticus* with *Rumex hydrolapathum* because he mistakenly assumed that the fruit of *Rumex aquaticus* could either bear tubercles or be smooth. He further suggested that "Linnaeus had already fallen into the same error ..." and that this was the reason why the water dock was not described and named in *Species Plantarum* – it had been included within *Rumex aquaticus*. However, it is difficult to believe that Linnaeus, in a short description to identify a plant, would seize upon a feature he knew to be variable.

*Fruits (with cross-sections)
of Rumex aquaticus (right),
R. longifolius (centre) and
R. hydrolapathum (left).*

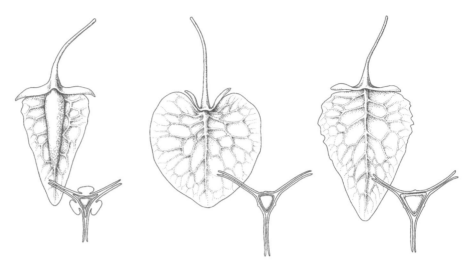

*Rumex aquaticus,
Stirlingshire, 7 viii 1995.*

Although William Hudson had provided the name of *R. hydrolapathum* for the water dock and Smith had stemmed the confusion with Linnaeus' *R. aquaticus*, further problems arose when William Jackson Hooker wrongly applied the name of *R. aquaticus* to yet another species, *Rumex longifolius*, what we now call northern dock. Even though the fruits of both these species lack tubercles, the shape of the fruits are much less acute in *R. longifolius* and the leaves are not triangular and heart-shaped at the base like *R. aquaticus* but broadly lance-shaped. Another useful distinction is the obvious ring-like joint on the fruit pedicel of *Rumex longifolius* but which is hardly discernible in *Rumex aquaticus*. True *Rumex aquaticus* had still not been found in Britain and to clear up this rather disconcerting muddle it urgently needed to be. In 1935 the very able Glasgow botanist Robert Mackechnie sent specimens of a puzzling *Rumex* that he found at Balmaha on the east side of Loch Lomond, to J. Edward Lousley. Lousley, an internationally recognized authority on docks, was able to confirm that Mackechnie had at last collected authentic *Rumex aquaticus* in Britain.

Since its discovery the Scottish dock has been located in scattered sites at the south end of Loch Lomond, especially beside the River Endrick and its tributaries. The plant occurs on river banks, wet meadows, clearings within alder woods and occasionally on sandy shingle. The main associates are branched bur-reed (*Sparganium erectum*), reed canary-grass (*Phalaris arundinacea*), bladder sedge (*Carex viscaria*), meadowsweet (*Filipendula ulmaria*), water horsetail (*Equisetum fluviatile*), common spike-rush (*Eleocharis palustris*) and tufted hair-grass (*Deschampsia cespitosa*). The requirement for permanent wetness could account for the late discovery of the plant as the flood plain of the River Endrick was drained in the 18th century and the habitat may have become too dry for *R. aquaticus*. When the drains fell into disuse the plant was able to gradually recolonize and build up the population.

As a species *Rumex aquaticus* has an almost circumpolar distribution but the Scottish plant, strictly *R. aquaticus* subsp. *aquaticus*, is European and western Asian. It reaches north to the southern tip of the Russian island of Novaya Zemlya at over 70°N and its southern and western outposts besides Scotland, are the Azores and north-west Spain. In central and east Asia the smaller subspecies *protractus* gradually replaces our subspecies, and other subspecies of the plant occur in North America.

The Scottish dock freely hybridizes with the broad-leaved dock (*Rumex obtusifolius*) and the hybrid (*Rumex* x *platyphyllos*), which can be detected by the teeth at the base of the fruit valves, is very common when the two species are growing in proximity. Two other much rarer *Rumex aquaticus* hybrids between it and the curled dock and wood dock (*Rumex crispus* and *Rumex sanguineus* var. *viridis* respectively) have also been found. Hybridization with *R. obtusifolius* is so frequent that there is considerable concern that this hybrid may eventually replace all pure *Rumex aquaticus*, especially if the habitat becomes drier for any reason. Attempts are being made by conservationists to keep *Rumex obtusifolius* away from *Rumex aquaticus* populations to prevent hybridization. Plant remains identified as *Rumex aquaticus* have been obtained from Late-glacial deposits in widely separated areas in Britain and this provides some evidence that the plant has been far more widespread and has become restricted during periods of drier climatic conditions.

Salix lanata L.
WOOLLY WILLOW

With its silver leaves and gold catkins, this plant is a rare jewel of the Scottish mountains. The first reliable record of *Salix lanata* in Britain is George Don's discovery in 1812 from Glen Callater, Aberdeenshire. Earlier, a willow resembling *S. lanata* was found by the Rev. Dr John Stuart at Finlarig by the south-west end of Loch Tay, which John Lightfoot reported in 1777. The plant soon found its way into the trade through Dr Stuart's garden at Luss, Loch Lomondside, and was widely distributed under the name *Salix lanata* var. *Stuartii*. Gardeners and botanists continued to debate the true identity of the plant, most suspecting it to be of hybrid origin. *S. lanata* x *S. lapponum* (downy willow) was the most favoured parentage, the same as for a willow described by the Rev. E. F. Linton from Glen Callater, south Aberdeenshire. Dr Stuart is still honoured in the scientific name for this hybrid, *S.* x *stuartii*.

Salix lanata in cultivation at Dawyck Botanic Garden, Peeblesshire, 12 vi 1995.

Salix lanata occurs in no more than 15 sites in the central Scottish Highlands between 600m and 900m. The westernmost site is in the Breadalbane mountains between Glen Lochay and Glen Dochart, and the furthest east is near Glen Clova in Angus.

One of the outstanding features of the British *Salix* flora, which numbers 18 native species, is the occurrence of several willows reaching their southern limit in Scotland. Woolly willow, mountain willow (*S. arbuscula*), whortle-leaved willow (*S. myrsinites*), downy willow (*S. lapponum*) and tea-leaved willow (*S. phylicifolia*) are all absent from central Europe, and outside Scotland are restricted to northern Europe and Russia. *S. lanata* occurs in Iceland, Faeröe, Scotland, Scandinavia and eastern Siberia. The Scottish Highlands has a flora most similar to southern Norway, an affinity highlighted by *S. lanata*, *S. arbuscula* and *S. myrsinites* which share a Northern Boreal distribution.

Montane willow scrub is a very restricted vegetation type in Scotland. It is found on ungrazed rocky slopes and ledges with wet, base-rich soils. Although probably a natural climax at high altitudes, it has been much reduced in extent and floristic diversity, primarily due to overgrazing. Communities containing *Salix lanata* are rarest of all, sometimes confined to a single cliff or ledge where sheep and deer dare not venture. Such restriction has in some cases isolated single male or female plants of *Salix lanata*, and this separation of the sexes greatly limits the reproductive potential of the species. In Glen Clova *S. lanata* is accompanied by *S. reticulata* and abundant *S. lapponum*, forming an impressive display. However, the close co-occurence of *S. lanata* and one or other willows facilitates the possibility of hybridization which could threaten small populations of pure *S. lanata*. Fortunately *S. lanata* does not appear to hybridize as readily as some of our lowland willows and only *S. herbacea* x *lanata* has been definitely recorded in recent years in Scotland.

Given a good supply of seed and suitable terrain, *S. lanata* can readily establish in the very early stages of vegetative succession. It is one of the first colonists of large glacier forefields in Norway, finding the newly-exposed moist, base-rich moraine an ideal substrate both for germination and vegetative spread by layering. It shows a strong preference for wet ground, and on the older, drier parts of the forefield it is strongly associated with stream channels, along which it forms dense thickets.

Salix lanata is well adapted to life above the tree-line. Its whitish pubescent leaves are highly reflective compared to the glabrous leaves of lowland willows, giving it an important advantage where air temperatures are relatively low due to strong winds and radiation is high in the rarified atmosphere. Inside a woolly willow canopy there exists a more sheltered microclimate, and the strongly pubescent leaves increase the amount of light reflected inwards to lower branches. A pubescent layer can act like the glass in a greenhouse, increasing the leaf temperature in a cold environment. The young catkins of woolly willow are also hairy, providing protection against damaging extremes of temperature during spring. The gradual loss of hairiness as the catkins mature coincides with a decreasing risk of lethal temperatures.

The catkins of many shrubs and trees such as hazel and birch are pendulous, facilitating pollen dispersal by wind. However, the catkins of most willows are erect and pollinated to a significant extent by bumble-bees which seek the nectar produced at the base of the individual flowers. Because they are early flowering and nectariferous, willows are highly prized by bee-keepers. One botanist/writer suggested *S. lanata* be sited near hives because it produces the most honey. Bees may provide a better chance than wind of carrying pollen between isolated male and female plants of *S. lanata*, although they may not discriminate between different willows while foraging.

Salix pollen is notoriously difficult to identify to species; macrofossils in the form of leaves being much more reliable evidence of a species past occurrence. Remains of *S. lanata*, *S. arbuscula* and *S. lapponum* have been identified from arctic-beds of Full-glacial age in southern England, evidence that these species once had a much wider distribution, which would have been favoured by the eradication of tree cover over most of Britain while the climate was much colder. There is no doubt that *S. lanata* has a very tenuous hold as a member of our present day flora. Besides deer damage which can be reduced or prevented, climate change is perhaps the most serious threat to its future.

Salix lapponum accompanies S. lanata at its Glen Clova locality.

93

Saxifraga aizoides L.
YELLOW SAXIFRAGE

Saxifraga hirculus L.
MARSH SAXIFRAGE

Yellow saxifrage was first recorded for Britain from Ingleborough and Shap in the Lake District by John Ray in 1670. The date of the earliest Scottish record is uncertain but the plant is listed, under the name of *Saxifraga autumnalis*, in the *Calendarium Florae* of John Hope in 1765 as being found "In low grounds at Denhead [= Deanhead, Dalkeith ?] and by rivulets in ye Highlands".

Saxifraga aizoides, Sutherland, 28 vii 1994.

Marsh saxifrage was discovered near Knutsford in Cheshire sometime between 1696 and 1720, and was included in the third edition of John Ray's *Synopsis* published in 1724. On Knutsford Moor the plant once grew in a bog with *Andromeda polifolia*, common cow-wheat (*Melampyrum pratense*), bog sedge (*Carex limosa*) and others. But in 1842 Dr J. B. Wood foretold the extinction of the plant in its most southerly locality; "[it] is almost destroyed by the rapacity of some individuals who have dug it up for sale in the most remorseless manner". In Scotland the earliest record is from Langton Lees Cleugh in Berwickshire in 1831. Unfortunately this locality was destroyed by drainage soon after 1886. A few years after the Berwickshire discovery, *S. hirculus* was found in the Pentland Hills, south-west of Edinburgh, where it still grows in springs near the source of the Medwin.

Saxifraga aizoides is neither rare nor scarce in Britain though it is decidedly Scottish. Only scattered occurrences on the north and north-west Irish coast and in north-west England extend its distribution beyond a concentration in the north and west of the Scottish mainland, with a northern outpost on Orkney. The present-day range of *S. hirculus* is highly fragmented, with sites in Aberdeenshire , Midlothian, the Lake District, and Counties Antrim and Mayo in Ireland. It is no longer known from several areas where it was recorded this century, namely Lanarkshire, Peeblesshire, west Perthshire and Caithness.

S. hirculus has declined or become extinct mainly through agricultural drainage over much of western Europe during the last two centuries. This increases the importance of the surviving British and Irish populations. Outside Europe the plant grows in arctic regions and in most of the major mountain ranges of Asia and North America. *S. aizoides* occurs in the mountains of central Europe and Scandinavia, across arctic and subarctic North America, but is absent from Asia except in the northern Urals and on the Russian island of Novaya Zemlya.

The yellow and marsh saxifrages are united not only by their flower colour, but in their confinement to wet habitats. Yellow saxifrage characteristically grows on the banks of mountain streams, on flushed grassy slopes or on rocky banks or cliff faces where there is a constant seepage of water. The plant's affinity for wet areas is noted on a specimen in John Hope's herbarium which reads "on all the rivulets in the hills in the North". But in addition to its demand for moisture, *S. aizoides* is also a marked calcicole. Where a yellow stripe or patch interrupts an otherwise brown hillside you can be sure that basic rocks or base-rich drainage water reaches the surface. In these golden carpets, *S. aizoides* is typically the dominant plant, but closer inspection will often reveal the rusty-golden moss, *Cratoneuron commutatum*, which enhances the distinctive yellow glow. On ungrazed crags and ledges *S. aizoides* grows in lush cushions of herbs which include alpine lady's-mantle (*Alchemilla alpina*), alpine meadow-rue (*Thalictrum alpinum*), purple saxifrage (*Saxifraga oppositifolia*), flea sedge (*Carex pulicaris*), and grass-of-Parnassus (*Parnassia palustris*), with bulky mosses and liverworts completing the matrix.

Saxifraga hirculus, Midlothian, 31 vii 1995.

S. hirculus is also found in mossy calcareous flushes but, being less of an alpine in Britain than *S. aizoides*, its primary habitat is not mountain sides but the edges of low-lying peat bogs where there is some movement of water to enrich the supply of nutrients. Its associates in this mesotrophic bog habitat include lesser spearwort (*Ranunculus flammula*), marsh lousewort (*Pedicularis palustris*), cuckooflower (*Cardamine pratensis*) and bottle sedge (*Carex rostrata*). It also occurs in shorter vegetation in flushes and vegetated springs dominated by mosses, small sedges and rushes with flowering plants such as knotted pearlwort (*Sagina nodosa*), ragged robin (*Lychnis flos-cuculi*) and marsh arrowgrass (*Triglochin palustris*).

Both *S. aizoides* and *S. hirculus* have efficient means of vegetative spread. The former has creeping stems which root to form new plants, while the latter produces runners which remain connected to the parent rhizome for one or more growing seasons before becoming separate shoots which form new rhizomes. Relatively few shoots of

S. hirculus produce a flower. Flower opening in this species is sensitive to temperature. In bright sunshine flowers open within a half to one hour, but in cloud opening is delayed by up to several hours. Flies are the most frequent visitors to the yellow flowers of both species, attracted by the red-orange spots at the base of the petals in *S. hirculus* (usually scattered in *S. aizoides*), which appear as a dark centre under reflected ultra-violet light. They consume the highly concentrated nectar as well as pollen which is also favoured by hover-flies. Rarer vistors, such as burnet moths, may be important for carrying pollen longer distances between different clones, and therefore ensuring cross-pollination.

Saxifraga hirculus, Midlothian, 31 vii 1995.

Seed production can be very high in *S. hirculus* but in closed vegetation seedlings are rare except following some disturbance which exposes seeds to high temperatures and light intensity, and provides bare ground for seedlings to establish. Bare ground is also required by *S. aizoides*, the seeds of which are transported by water and probably birds.

Saxifraga cernua L.
DROOPING SAXIFRAGE

Drooping saxifrage was first found in Britain by Robert Townson in 1790 on Ben Lawers, as recorded by William Hooker in *Flora Scotica* in 1821. Hooker wrote, "I am not aware that it has been detected anywhere else, but upon Craigalleach [Creag na Caillich], by Mr. Borrer and Hook". By 1898 the plant was reported to have disappeared from this second Breadalbane mountain and Ben Lawers remained its only known site until 1949 when it was found on Bidean nam Bian in Glen Coe. A third site was discovered in 1956 on the Ben Nevis range. In these localities the plant grows not far below the summit of what are three of Scotland's highest mountains.

Worldwide the species has a primarily arctic distribution, but its total range is extended southwards by montane occurrences in Scotland, the European Alps, the Kjølen Mountains of Norway, the Himalaya and Siberia, central Japan and the Rocky Mountains of North America.

The alpine habitat of *S. cernua* is typically rocky, and always moist. The plant finds necessary shelter and moisture at the base of rock overhangs, on cliff ledges or in gullies where melt-water from snow beds provides irrigation. It shows a preference for

96

base-rich sites, and amongst its associates in Scotland are other calcicolous saxifrages. In the Arctic, *Saxifraga cernua* grows in similarly damp and cold localities, in wet tundra, in moss cushions, along stream banks and on stones in rivers. The plant has no means of protection against excessive water loss, hence its intolerance of drier and warmer sites. The kidney-shaped leaves are hairless, with raised pores, a thin cuticle and large intercellular spaces, all features which allow water to be lost from the leaves relatively easily.

Saxifraga cernua has attracted the attention of botanists because of its rarity in most of its sub-arctic stations, and its ability to survive the most extreme climatic conditions. In all the sub-arctic populations investigated, the plant survives despite failing to reproduce sexually. Fertile pollen, and hence seed, appears to be produced only in populations north of about 66°N. Plants in the Scottish populations have diminished pollen fertility and are virtually male sterile. However, they retain the capacity for seed formation, as shown when plants on Ben Lawers were hand-pollinated with pollen from plants in arctic Norway. In the absence of sexual reproduction the plant relies entirely on vegetative propagation by means of small red bulbils produced in the axils of the leaves. These bulbils can start to germinate while still on the plant, but usually they first detach from the stem and disperse a short distance away from the parent plant. Occasionally, bulbils may be more widely transported, especially by strong winds which buffet mountainsides and sweep across the arctic tundra.

Flowering is generally more common in the Arctic than in southern alpine populations, but it also varies between habitats. Plants in the Glen Coe site seem to be less shy than those on Ben Lawers.

The name *cernua* means drooping or nodding, and was given to the plant by Linnaeus in reference to the flower, or the flower bud. The usually solitary flower opens from June onwards and may be visited by flies, but in the absence of pollen, no seed is produced. The Scottish populations thus probably consist of a single clone (i.e. all plants are genetically identical).

The same situation can prevail locally in the Arctic due to the high efficiency of bulbil production. In any one population new plants that germinate from bulbils may greatly outnumber those produced from seed. However, because pollen can be carried between clones and seed is dispersed much farther than bulbils, genetic variation is maintained despite a relatively low level of sexual reproduction. These populations therefore retain the potential for evolutionary adaptation which may be lost in areas where only vegetative propagation occurs.

Saxifraga cernua, Perthshire, 13 vii 1995.

Saxifraga oppositifolia L.

PURPLE SAXIFRAGE

Saxifraga oppositifolia heralds the coming of spring in our mountains. It is by far the earliest montane plant to flower and is probably enjoyed more by hill-walkers than botanists.

Although much more widespread in Scotland than in England, the plant was first discovered in Yorkshire "On the north side of the summit of Ingleborough" by John Ray in 1668. This reflects the earlier botanical exploration of the English hills.

Saxifraga oppositifolia,
Perthshire, 19 iv 1995.

S. oppositifolia has a northern and western tendency in Scotland but occurs from the Southern Uplands, throughout the Highlands to the Inner and Outer Hebrides and the Northern Isles of Orkney and Shetland. Elsewhere in Britain, because of the more restricted montane habitat, *S. oppositifolia* is confined to the mountains of the Lake District, the northern Pennines in Yorkshire, and Snowdonia and the Brecknock Beacons in Wales. In Ireland its scattered localities are mainly in the west and north-west.

S. oppositifolia is such a reliable indicator of basic rocks in Britain that it can be useful in the detection of small calcareous pockets on otherwise acid mountains. The plant occurs most commonly on mountain ledges at medium to high altitudes and reaches 1210m on Ben Lawers but descends to near sea-level on the coasts of Aberdeenshire and west Sutherland. Towards the south of its range in Britain *S. oppositifolia* is confined to cooler, northerly aspects, but in Scotland it is not so restricted and often occurs on sunny slopes although overhanging rocks may sometimes protect the plant from direct sun.

Even though *Saxifraga oppositifolia* can be locally abundant it rarely attains dominance within its associated vegetation. It is usually present in the ungrazed mountain ledge community where mountain avens (*Dryas octopetala*) and moss campion (*Silene acaulis*) are conspicuous. However, as the plant is usually prostrate, it escapes grazing and therefore also grows in a similar, but more grazed community where *Dryas* is replaced by a greater abundance of the more resilient alpine lady's-mantle (*Alchemilla alpina*). The tolerance of purple saxifrage to wet conditions enables it to be prominent in dripping banks of vegetation where yellow saxifrage (*Saxifraga aizoides*) dominates and mossy saxifrage (*S. hypnoides*) can be common. As *S. oppositifolia* is less restricted to high altitudes than many other montane plants, it may be found on lower slopes in calcareous grassland with fewer of these species.

The global distribution of *Saxifraga oppositifolia* is circumpolar and it is one of the hardiest arctic-flowering plants. Indeed it vies for the accolade of being the most northerly reaching, having been recorded from the north coast of Greenland at Cape Morris Jessup (83°39′N). It is also scattered through the southern European mountain ranges and in North America it reaches south to Newfoundland.

The plant is also confined to base-rich rocks in Scandinavia, but in other parts of its geographical range, especially in most of the Alps and arctic Canada, it is less marked in its soil preferences. It is extremely frost hardy and does not require winter snow protection even at high latitudes, yet it also grows in exceptionally long-lasting snow patches in the Arctic.

Saxifraga oppositifolia is a variable species over its wide geographic range and several subspecies and varieties have been distinguished. The main trends are variation in size and colour of petals (deep purple through pink to white), the shape and arrangement of the leaves, and the structure and distribution of hairs on the leaves and sepals. One of the most interesting variants is aptly named subsp. *paradoxa* because its leaves do not exhibit the characteristic opposite arrangement but are placed alternately. Especially attractive colour forms and those with large flowers have been selected and given horticultural names.

Saxifraga oppositifolia,
Perthshire, 26 iv 1994.

To survive in areas with a very short-growing season demands adaptation. As with many other species in the Arctic, the increase in temperature following snowmelt triggers flowering in *S. oppositifolia*, but in the Alps it may flower beneath the snow, so daylength may override snowmelt as long as the temperature is above a critical level. Flowers overwinter in an advanced state of development and as the plant resumes growth under very low temperatures early flowering occurs. The prostrate shoots obtain maximum benefit from heat reflected from the rock or soil surface and the bowl-shaped flowers act as heat traps to help hasten seed development. *S. oppositifolia* may form either cushions or trailing mats depending on the degree of branching of the main stems. In Scotland both forms have been seen growing almost side-by-side with little evidence of a difference in habitat. In the high arctic there is a more marked separation with the cushion forms growing on raised dry ridges whilst the prostrate mats grow on wet, late-thawing shores. The cushion-formers are slow growing and accumulate food reserves for drought conditions in summer, whilst the trailing mats grow rapidly in the even shorter growing season of wetter habitats.

The flowers produce abundant nectar and depend on cross-fertilization for maximum seed-set. In Scotland the only pollinator recorded is a blow-fly, but in the Arctic bees, flies and butterflies have been observed. Reduced seed production generally occurs in self-pollinated flowers but is variable between populations. Seed production is mostly limited by lack of pollinators but occasional years of little or no seed production are probably not crucial for the long-term maintenance of the population as individuals are very long-lived.

Silene acaulis (L.) Jacq.

MOSS CAMPION

Anywhere in the richer mountains in the Scottish Highlands one is likely to encounter, in June, the attractive pink-flowered mats or cushions of moss campion. The plant is a useful indicator of promising areas for the mountain botanist and quite often it is passed by because concentration is focused on what else might be found in its vicinity!

Largely due to the very early pioneering exploration of the Welsh mountains by Edward Lhwyd, many montane plants that occur in Wales and Scotland were first discovered in Wales. John Ray first reported *Silene acaulis* from Snowdon in the first edition of his *Synopsis Methodica Britannicarum* in 1690 but the record came from Lhwyd's catalogue which was sent to Ray by his "worthy-friend" Dr Tancred Robinson.

Apart from its very restricted occurrence in North Wales, the Lake District and Northern Ireland, *Silene acaulis* is very much a plant of the Scottish Highlands although it extends to the Inner and Outer Hebrides and the Northern Isles of Orkney and Shetland. It reaches its highest altitude in Britain near the summit of Ben Macdhui at 1309m in the Cairngorm mountains and descends to near sea level in west Sutherland, the Western Isles and in Shetland.

Moss campion usually grows on ledges or in crevices but also occurs on more level ground on mountain plateaux and, rarely, in stabilized sand dunes in the Western Isles. At higher elevations it occurs on a wide range of rocks from acid to basic but lower down seems to be more restricted to basic rocks and is generally more abundant on this rock type.

In its montane habitat *Silene acaulis* occurs in a range of plant communities but is most prominent and constant in high altitude vegetation. Here, with sheep's fescue (*Festuca ovina*), wild thyme (*Thymus polytrichus* subsp. *britannicus*), alpine lady's-mantle (*Alchemilla alpina*) and cushion-forming plants such as cyphel (*Minuartia sedoides*) and sibbaldia (*Sibbaldia procumbens*), it characterizes this dwarf-herb-dominated vegetation. Several montane rarities may also occur in this vegetation type. On less accessible ledges, freer from grazing, mountain avens (*Dryas octopetala*) may form a distinctive ledge community with *Silene acaulis* where, as well as some associates of the dwarf-herb vegetation, some of the attractive montane willows may be prominent. It has been suggested that the *Silene acaulis* dwarf-herb community may be derived from the *Dryas octopetala-Silene acaulis* community through grazing. As isolated plants, *Silene acaulis* extends into snowbed vegetation, and also in the sparse vegetation on exposed summits. Moss campion is essentially a plant of cold, wet and windy places and its tolerance of acid soils at high altitudes is likely to be due to the redistribution of available nutrients through frost-churning. Tolerance to extremely harsh conditions is reflected in the global distribution of *Silene acaulis*, reaching 83°9′N in Greenland and, up to an altitude of 3700m in the Alps. It is also one of the highest ascending alpine plants.

Silene acaulis, Perthshire, 7 vi 1994.

The efficient absorption and retention of heat by the cushions of *Silene acaulis* allow the plant to survive extremely cold conditions. Further heat retention is afforded by the internal structure of the leaves which have large air spaces on the upper sides. Although the structure of the plant equips it well for cold environments, this becomes limiting in its more southerly locations. Consequently in Wales, for example, it is confined to moist, steep north- and east-facing slopes.

The Scottish plant is strictly *Silene acaulis* subsp. *acaulis* and is the only subspecies which occurs in Britain, but in the Alps subsp. *bryoides* also occurs which is more restricted to acid rocks. Subsp. *bryoides* and a further subspecies, are the only forms in North America. The species as a whole is widely distributed in montane and arctic regions but with a wide gap in Siberia. In Europe it reaches south to the Apennines.

Silene acaulis is well adapted for cross-pollination. The bright pink flowers are visually attractive to a range of insects but moths and butterflies are the main visitors. The male, female or hermaphrodite flowers may be confined to separate plants thus displaying an unusual condition known as trioecy, but more commonly the flowers are mixed in various combinations on individual plants. As the anthers of hermaphrodite flowers mature in advance of the stigmas, it is sometimes difficult to appreciate that they are bisexual. In Scotland flowering usually occurs from June to July but on the exposed serpentinite hill of the Keen of Hamar in Shetland, the markedly convex cushions flower in May and are withered by the time the flowers of the Shetland mouse-ear open in June.

In the Arctic, the flowers of *Silene acaulis* have been observed to open sequentially, first on the south side gradually moving across the cushion to the north by which time the flowers on the south side are producing fruits. Because of this it is feasible that the plant may be used for navigation if one is lost in the arctic wastes, but dependence on this to the exclusion of a compass cannot be recommended!

The flower colour varies from white through pink to deep rose and double flowers are not rare. The great Yorkshire rock gardener Reginald Farrer considered the plant over-rated and thought that however brilliant the shade of pink it was never "clean". In comparison with the pure pink mats of *Androsace alpina* which grows in the vicinity of *Silene acaulis* in the Alps his feelings were that "... one sickens evermore of the *Silene*, despite one's best intentions and honest feelings of admiration."

Sorbus arranensis Hedl.
ARRAN WHITEBEAM

Sorbus pseudofennica Warburg
ARRAN SERVICE-TREE

In the north temperate regions of the world the genus *Sorbus* contains over 100 species which are widely valued for the beauty of their fruit and foliage. In Britain there are three widespread species which reproduce by normal sexual fertilization: common whitebeam (*Sorbus aria*), wild service-tree (*Sorbus torminalis*) and rowan (*Sorbus aucuparia*) and of these, rowan is the only one native to Scotland. Besides these sexual species, there are many that have arisen by multiplication of chromosome sets, hybridization or a combination of the two. Linked to this is replacement of normal sexual reproduction with the ability to produce fertile seed without fertilization (agamospermy). There are several variations in the process of agamospermy and in some cases pollination is not necessary to trigger seed formation but in *Sorbus* this must occur for successful seed-set. The pollen often comes from one of the sexual species as the pollen viability of the agamospermous species is reduced. However, as no sexual fusion takes place the pollen donor makes no genetic contribution to the offspring. This method of reproduction is non-sexual but, unlike other methods such as bulbs, runners or off-sets cannot be considered vegetative because the floral organs are involved. For this reason the general term apomixis is used to embrace all these methods of non-sexual reproduction.

There are 15 agamospermous species of *Sorbus* which are considered endemic to the British Isles. Many have a very restricted distribution and two, the Arran whitebeam (*Sorbus arranensis*) and the Arran service-tree (*Sorbus pseudofennica*) are, as their names suggest, endemic to the island of Arran.

The exact dates of discovery of these rare Arran sorbi are not known, but *S. pseudofennica* appeared in the first edition of James Sowerby's *English Botany* under the name *Pyrus pinnatifida*; its inclusion in this work was based on specimens that Sowerby had received from James Mackay in 1797 gathered "... on rocky places on Cairn y caillich [?] (sic) and other mountains, north end of the isle of Arran".

The earliest known herbarium specimens of *Sorbus arranensis* are those of Alexander Craig-Christie dated 1869. For a considerable time after the discovery of the Arran sorbi there was considerable debate among botanists over their identity and they were usually identified with similar Scandinavian species. Even when Johan Teodor Hedlund, in a monograph of the genus, considered the Arran whitebeam a new species and named it *S. arranensis* Hedl., he did not consider it endemic to Arran but "... widely distributed in Norway". However, some botanists were already of the opinion that the plant was endemic to Arran. General acceptance of its endemism by British botanists seems only to have occurred since E. F. Warburg's account in the first edition of *Flora of the the British Isles* in 1952. It was also in this work that Warburg considered the Arran service-tree as separate from the Scandinavian species with which it had previously been identified (*S. fennica* or *S. hybrida*) and named it *Sorbus pseudofennica* Warburg.

Sorbus arranensis in cultivation at the Royal Botanic Garden Edinburgh, 12 vi 1995.

Both species are restricted to the north end of Arran and occur in rocky steep-sided glens from about 90 to 300m. *S. pseudofennica* is the more restricted of the two, only occurring in the vicinity of Glen Diomhan and Glen Catacol where it accompanies *S. arranensis*. The latter is known from the environs of at least two other glens in north Arran. Glen Diomhan was declared a National Nature Reserve in 1954 to protect these rare trees through appropriate habitat management. The present restriction in their occurrence is thought to be a result of contraction and fragmentation of their range, primarily through forest clearance for agriculture followed by prevention of regeneration by grazing animals, mainly sheep, deer, and hares. Their confinement to rocky sites renders the two *Sorbus* species vulnerable to gales and snow which can dislodge mature specimens. The associated flora is sparse on the steep rocks but rowan (*Sorbus aucuparia*), downy birch (*Betula pubescens*), juniper (*Juniperus communis*), honeysuckle (*Lonicera periclymenum*) and bearberry (*Arctostaphylos uva-ursi*) have all been recorded. Unlike other British agamospermous sorbi, *S. arranensis* and *S. pseudofennica* grow on acid granite rather than calcareous rocks. However, there is evidence that the plants are associated with areas of soil enrichment caused by the occurrence of calcium-rich minerals (epidotes) which have formed along the joints of the granite rocks. They also seem to occur in the vicinity of dykes of other basic minerals. Neither of the two sorbi are tolerant of organic peaty soil and it is thought that the failure of a trial reinforcement programme in the early 1980s was due to planting young saplings on organic rather than mineral soil.

Sorbus pseudofennica in cultivation at Dawyck Botanic Garden, Peeblesshire, 5 x 1994.

From chromosome studies and various characters of the leaves it is generally accepted that *Sorbus arranensis* is of hybrid origin resulting from a rare cross between rock whitebeam (*Sorbus rupicola*) and rowan (*Sorbus aucuparia*). Genetic copies of *S. arranensis* could be produced by agamospermy. On other very rare occasions it is assumed that *S. arranensis* has reproduced sexually and backcrossed to one of its parents (*S. aucuparia*) to form *S. pseudofennica*. In other geographical areas of Britain and Scandinavia there are other recognized species that are thought to have the same parentage as *S. arranensis* and *S. pseudofennica* but differ in their morphology (mainly in details of leaf and fruit shape). This probably reflects the natural genetic variation of the parent species. The two British sorbi considered to have the same parentage as *S. arranensis* are *S. leyana* and *S. minima* (both restricted to Brecon), whilst *S. anglica*, which occurs in south-west England, Wales and Ireland, shares the same parentage as *S. pseudofennica*.

Not surprisingly it is sometimes difficult to distinguish between individuals of *Sorbus arranensis* and *S. pseudofennica*. Generally the leaves of *S. arranensis*, although deeply incised, do not possess free leaflets at their base whilst many of those of *S. pseudofennica* (but by no means all on the same tree) have one to two pairs of free leaflets at the base, reflecting its closer association with *S. aucuparia*. A most interesting specimen in the herbarium of the Royal Botanic Garden Edinburgh, collected by Dr Donald McVean, has leaves which possess four pairs of free leaflets giving the strong impression that it may be a hybrid between *S. pseudofennica* and *S. aucuparia*.

A detailed study of the two rare Arran sorbi has shown that the degree of variation between populations of *S. arranensis* is significantly greater than that of *S. pseudofennica*. The greater interpopulation variation of *S. arranensis* could be explained either by the formation of hybrids involving different parental individuals or by mutation occurring within isolated populations. This is, on a local scale, the same phenomenon that has given rise to what are considered separate species in other geographical locations (*S. minima* and *S. leyana* for example). *S. pseudofennica* is considered to have retained a greater propensity for sexual reproduction than *S. arranensis* resulting in increased genetic exchange between populations. This has resulted in greater variation among individuals than *S. arranensis*, but spread throughout the total Arran population, rather than the more marked differences between populations in the latter species.

Leaves of Sorbus rupicola (left),
S. aucuparia (centre left),
S. arranensis (centre right) and
S. pseudofennica (right).

Though the genus *Sorbus* contains several apomictic species, there are far fewer than in some other British genera and they are mostly separated geographically and reasonably distinguishable. For this reason there are fewer conservation problems than with other groups such as hawkweeds (*Hieracium*) and brambles (*Rubus*) where several hundred species have been named.

Even taking into account the fact that one of the parents of *Sorbus arranensis* and *S. pseudofennica*, *S. rupicola*, is much scarcer than the other, *S. aucuparia*, both are quite widespread. The rarity of the hybrid apomictic species is therefore an indicator of the rarity of this biological event and thus is of considerable intrinsic interest.

Vaccinium microcarpum
(Turcz. ex Rupr.) Schmalh.
SMALL CRANBERRY

Vaccinium oxycoccus L.
CRANBERRY

The first mention of cranberry in Britain seems to have been by Henry Lyte in his herbal of 1578, but John Gerard was the first to record a specific locality, " ... upon bogs and fennie places especially in Cheshire and Staffordshire". This was for *Vaccinium oxycoccus*, not *V. microcarpum* which has only been recorded from Scotland.

V. microcarpum is considered a scarce plant in Britain although its true distribution is not known because of the difficulty in distinguishing it from *V. oxycoccus*. *V. oxycoccus* is certainly more widespread, its range stretching the length of the

Habitat of
Vaccinium microcarpum,
East Ross, 21 vi 1995.

country, but with notable concentrations in southern Scotland, north and north-west England and west Wales. *V. microcarpum*, on the other hand, is more or less restricted to Caithness and Grampian with scattered occurrences further south and west. Its range has been extended in recent years by the addition of new records, and also as a result of the reidentification of herbarium specimens originally determined as *V. oxycoccus*. This has inevitably led to the 'resurrection' of certain localities, such as the Isle of Skye, where it was first recorded in 1868 (as *V. oxycoccus*) and not again until 1981. However, with the destruction of many lowland raised bogs, both species have lost sites this century, especially *V. oxycoccus* in the south of Britain.

Within the subgenus *Oxycoccus* of *Vaccinium*, two entities were originally recognized, *V. oxycoccus* and *V. macrocarpon*. The latter, the commercial cranberry of North America is present in Britain only as an introduction. Taxonomists have split *V. oxycoccus* (in the wide sense) into two (or more) species based on the fact that there are plants with different chromosome numbers; *V. oxycoccus* (in the strict sense) has twice the chromosome number of *V. microcarpum*. However, distinguishing morphological characters linked to plants with different chromosome numbers have been hard to find. As a general rule, *V. microcarpum* has hairless flower stalks and very small, triangular inrolled leaves, whereas *V. oxycoccus* has hairy flower stalks and larger, parallel-sided leaves. Confusing intermediates occur within populations and hybrids between the two species are known.

The cranberries are circumboreal, extending from the low arctic tundra in northern Europe and North America to the European Alps and Japan at their southern limits. Despite having very similar distributions, *V. microcarpum* and *V. oxycoccus* differ ecologically. However, where they do overlap, the mosaic nature of their habitat means that populations of the two species can be found in very close association. Both cranberries are exclusive to mires, but *V. oxycoccus* grows in a greater range of mire communities, from floating quagmires to relatively dry wooded areas. It is also somewhat more sprawling in habit, forming more extensive colonies than *V. microcarpum*. The wider environmental tolerance and growth behaviour of *V. oxycoccus* is typical of many species which have doubled their chromosome number (polyploids). *V. microcarpum* tends to occur at higher altitudes in Europe, including Scotland, but the only place where it really replaces *V. oxycoccus* is in Lapland.

Like *Andromeda polifolia*, the cranberries are characteristic of raised bogs. The raised bog ecosystem represents a rich mosaic of habitats and communities on a very small scale; this spatial diversity is produced by the permutation of around 30 higher plants and a similar number of mosses and lichens. Only about a third of these species are plentiful, so rigorous is the selection in this extreme habitat.

V. oxycoccus is intolerant of constant submersion and tends to colonize watery hollows only once they have begun to fill up from the bottom with sphagna, notably the yellow-green *Sphagnum cuspidatum*. In general, dwarf shrubs find only a hesitant footing in the transition zone between hummock and hollow; among the more successful are *V. oxycoccus* and *Andromeda polifolia*.

V. microcarpum is less tolerant of a saturated root run and generally grows on drier ground, either on top of moss hummocks or in the relatively dry micro-habitat provided by tussocks of cotton grass (*Eriophorum vaginatum*) or heather (*Calluna vulgaris*).

Vaccinium microcarpum, East Ross, 22 vi 1995.

Any vascular plant growing in bog pools or on their margins can become engulfed by the accumulating peat and *Sphagnum* mosses. The long shoots of cranberry would be buried by *Sphagnum* overgrowth within two years if they did not extend continually to keep pace with the changing levels of the bog surface. The thin wiry stems of cranberry are superbly able to grow up through the shoots of the moss and creep above their tips.

The xeromorphic character of bog plants which long puzzled plant physiologists and ecologists, is a reflection of the extremely poor supply of nutrients. Thus it is not true xeromorphism (morphological adaptation to drought), but peinomorphism, literally 'hunger structure'. The cranberries are typically peinomorphic. Their leaves are not only very small and hard, they are covered beneath by wax. In *V. microcarpum* the wax is so thick it obscures the stomata, whereas in *V. oxycoccus* some stomata are still visible. This difference agrees with the general preference of small cranberry for drier microsites. The leaves are also evergreen, their longevity enabling the plants to use the mineral nutrients acquired to make them over a number of years. Their peinomorphic structure allows copious transpiration, and to some extent the large volume of water (containing dilute nutrients), passing through the plant helps compensate for the very low nutrient concentration of the bog water. Proof of the importance of the poor nutrient supply is that in plants of *V. oxycoccus* fertilized with nitrogen, the leaf structure becomes more mesomorphic, the cells grow larger and the density of leaf pores is reduced.

The cranberries' boggy habitat is reflected in several common names: mossberries, fen grapes, marsh whorts

and the Gaelic, muileag, meaning a little frog. There are two explanations for the original English name of craneberry. One is that the fruit ripens (or remains on the plant) in spring, heralding the return of the crane. The other is that the flower with its crest of pink 'plumes' and sharp, down-pointing beak of stamens resembles the head of this bird. The name oxycoccus is from the Greek *oxys*, sharp and *kokkos*, berry from the acid-tasting fruit.

The delicate flowers are perhaps more like those of cyclamen or a miniature turk's cap lily than a crane's head. They are pollinated by bumble-bees seeking nectar, though insects are rarely seen. The fruits are borne on such slender stalks that as they fill they droop onto the surface of the moss where some may lie dormant all winter, inconspicuous against the reds

Vaccinium oxycoccus, Stirlingshire, 6 vi 1992.

and pinks of the sphagna. The berries are mostly eaten and dispersed by birds such as grouse, blackcock and the rare capercailzie. Sweetened with sugar for the human palate, they are delicious in tarts and restore an appetite to any meat in cranberry sauce.

Woodsia alpina (Bolton) S.F. Gray
ALPINE WOODSIA

Woodsia ilvensis (L.) R. Br.
OBLONG WOODSIA

The earliest record of a *Woodsia* fern in Britain is in John Ray's *Synopsis* of 1690 in which Ray reported Edward Lhwyd's discovery of this "very rare plant" on "Clogwyn y Garnedh", near the top of Snowdon. Ray's Latin synonym translates as "stone-fern with red-rattle leaves, hairy underneath", which to botanists suggests *Woodsia ilvensis*. However, specimens collected by Lhywyd from Snowdon have been determined as *Woodsia alpina*. Not until the publication of James Bolton's *Filices*

Britannica in 1785 was it realised that two fern species had been confounded under the name *Acrostichum ilvense*. Bolton described a second species, *Acrostichum alpinum* (= *Woodsia alpina*) from specimens collected in Scotland. Thus alpine woodsia can be said to have had a Scottish birth.

The genus *Woodsia* was established in 1810 by Robert Brown, honouring fellow British botanist, Joseph Woods, who is remembered for his *Tourist flora of the British Islands* of 1850. Practical confusion has continued in identifying the two woodsias. The most reliable differences are the scales and many hairs on the fronds of *W. ilvensis*, whereas *W. alpina* has few or none, and the greater degree of lobing of the former's leaf segments. Close morphological similarity arises from the relationship of the two species; *W. alpina* is the chromosome-doubled hybrid between *W. glabella* (absent from Britain) and *W. ilvensis*. Because of identification difficulties there is much uncertainty about the identity of many early *Woodsia* records.

Alpine and oblong woodsia share some common ground in Britain; both species occur in north Wales and in Angus. *W. ilvensis* is additionally known from the Lake District, Teesdale, Angus and the Moffat Hills, whereas *W. alpina* occurs in several sites in Perthshire and Argyll. True to its name, *W. alpina* is more strictly alpine, being found only above the potential tree-line (about 600m) up to 950m, while *W. ilvensis* is more sub-alpine, growing between 350m and 720m in Britain.

Both woodsias are more or less circumpolar, but *W. alpina* is considerably more scattered across Asia and is only really abundant in the mountains of Scandinavia. It also occurs near sea level on the Swedish island of Öland where it is regarded as an alpine relict along with several other species unexpected this far south: alpine meadow-grass (*Poa alpina*), holly fern (*Polystichum lonchitis*) and alpine catchfly (*Lychnis alpina*). *W. ilvensis* has a similar, but larger concentration in Scandinavia, extending round the Baltic, in addition to being especially abundant in the Urals, Altai and the eastern United States.

Woodsias are light-loving ferns of open rock crevices, cliff terraces and scree slopes. Such sites are very free-draining with little soil and are consequently strongly influenced by the type of rock. The two species have been recorded on a range of rock types, including basalt, pumice tuffs, silurian grits and slaty shales. Both species are moderately basiphilous, increasingly so in the north of their range. In Britain *W. alpina* seems the more demanding of base-rich substrates, whereas *W. ilvensis* predominates on slightly acid to neutral sites. The westerly distribution of both species reflects their intolerance of high summer temperatures and a requirement for high rainfall, though in their rocky microsites plants are often sheltered from the worst of heavy rain and winds which ravage exposed mountainsides. The ferns usually fare best in wetter summers and can take several years to recover from severe droughts.

Woodsias spread by creeping rhizomes which send up crowns through rock fissures, and also by spores released from fertile fronds in late summer. By these two methods of propagation quite large local colonies can be formed. However, some British *Woodsia* colonies are so small that their viability is in doubt.

Woodsia alpina, Perthshire, 27 vii 1995.

There is overwhelming evidence that woodsias suffered catastrophic decline due to overcollecting during several decades of Pteridomania or 'fern fever' which gripped Victorian Britain in the mid-19th century. Collecting led to the near extinction of *Woodsia ilvensis* in the Moffat Hills, a saga comprehensively catalogued by Scottish botanist, John Mitchell. The completion of the Caledonian Railway in 1848 marked the beginning of this onslaught, where visiting botanists, nurserymen, specialist fern dealers, Moffat hill-shepherds and the Innerleithen Alpine Club were involved. The precipitous nature of many sites proved an insufficient deterrant, rather it probably increased the attraction of the prize, though not always without cost. One particularly treacherous cliff nearly cost John Sadler, future curator of the Royal Botanic Garden Edinburgh, his life as he attempted to reach a fern tuft. It is believed that William Williams, a botanical guide on Snowdon, perished in 1861 while collecting specimens of *Woodsia alpina* for his clients. Williams' body was found at the foot of the very same precipice where Edward Lhwyd first collected this fern.

In the Moffat Hills, once the British headquarters for *W. ilvensis*, there now remains a handful of very small colonies whose future is uncertain, a plight echoed by several other surviving sites of both woodsias. However, important stocks of these and other endangered ferns are now being kept in cultivation and in a spore bank at the Royal Botanic Garden Edinburgh. Some of these will be used in attempts to help depleted wild populations recover.

While the blame has been squarely laid on the Victorians for the sudden demise of the woodsias, from a longer historical perspective such persecution has probably been only a final blow in a protracted period of natural decline. Both woodsias are considered to be on the edge of their ecological ranges in Britain and, as such, existing in a delicate balance with their environment. The woodsias were primarily sought after *because* of their rarity, not in spite of it. However, since their reduction to even smaller and fewer populations they have become more vulnerable to episodes of unfavourable weather and to the more worrying trend of climatic warming.

Woodsia ilvensis, Cumbria,
15 ix 1995.

Index

The Stationery Office

Published by The Stationery Office and available from:

The Stationery Office Bookshops
71 Lothian Road, Edinburgh EH3 9AZ
(counter service only)
South Gyle Crescent, Edinburgh EH12 9EB
(mail, fax and telephone orders only)
0131-479 3141 Fax 0131-479 3142
49 High Holborn, London WC1V 6HB
(counter service and fax orders only)
Fax 0171-831 1326
68-69 Bull Street, Birmingham B4 6AD
0121-236 9696 Fax 0121-236 9699
33 Wine Street, Bristol BS1 2BQ
0117-926 4306 Fax 0117-929 4515
9-21 Princess Street, Manchester M60 8AS
0161-834 7201 Fax 0161-833 0634
16 Arthur Street, Belfast BT1 4GD
01232 238451 Fax 01232 235401
The Stationery Office Oriel Bookshop
The Friary, Cardiff CF1 4AA
01222 395548 Fax 01222 384347

The Stationery Office publications are also available from:

The Publications Centre
(mail, telephone and fax orders only)
PO Box 276, London SW8 5DT
General enquiries 0171-873 0011
Telephone orders 0171-873 9090
Fax orders 0171-873 8200

Accredited Agents
(see Yellow Pages)

and through good booksellers